# A Primer of Mathematical Writing

Being a Disquisition on Having Your Ideas Recorded, Typeset, Published, Read, and Appreciated

Second Edition

Steven G. Krantz

AMERICAN MATHEMATICAL SOCIETY
Providence, Rhode Island

2010 *Mathematics Subject Classification.* Primary 00-01, 00-02.

For additional information and updates on this book, visit
**www.ams.org/bookpages/mbk-112**

**Library of Congress Cataloging-in-Publication Data**
Names: Krantz, Steven G. (Steven George), 1951- author.
Title: A primer of mathematical writing: being a disquisition on having your ideas recorded,
   typeset, published, read, and appreciated / Steven G. Krantz.
Description: Second edition. — Providence, Rhode Island: American Mathematical Society,
   [2017] — Includes bibliographical references and index.
Identifiers: LCCN 2017028388 — ISBN 9781470436582 (alk. paper)
Subjects: LCSH: Mathematics–Authorship. — AMS: General – Instructional exposition (text-
   books, tutorial papers, etc.). msc — General – Research exposition (monographs, survey
   articles). msc
Classification: LCC QA42 .K73 2017 — DDC 808.06/651–dc23
LC record available at https://lccn.loc.gov/2017028388

This book is dedicated to Paul Halmos.
For the example he has set for us all.

# Contents

# Preface to the Second Edition

The reader response to the first edition of this book has been gratifying. Especially because of the Internet, and the cognate rapid and free dissemination of scholarly work, people are now paying more attention to the quality of writing. And we are all benefiting from the result.

The essential principles of good writing have not changed for many years. In this new edition, I am not going to revise my advice about grammar and syntax and organization and style. I will certainly update and amend and correct certain passages. But the basic message will be much as in the first edition.

I will still insist that writing is a yoga, and a healthy one. It is a discipline that one must cultivate in oneself, and it is one worth cultivating. In today's world, good writers are respected and admired. They are granted a considerable measure of accord and prestige. They are a crucial part of our discipline.

What will be truly new in this second edition is an extensive discussion of technological developments. Today the Internet virtually consumes all of our lives (and especially of the lives of writers). As both readers and writers, we are all aware of blogs and chat rooms and preprint servers. There are now electronic-only journals and print-on-demand books and Open Access journals and joint research projects such as `MathOverflow`. Not to mention a host of other new realities. It truly is a brave new world, one that can be overwhelming and confusing.

Put slightly differently, information-and-misinformation are today vastly more accessible to everyone than they were yesterday. This basic fact has altered, and largely benefited, communication, education, and scholarship. It has redefined, perhaps in some sense toppled, the Ivory Tower. And it has heightened our responsibility to be accurate, responsible, and concise communicators of our science.

With some trepidation, I have attempted here to describe and catalog this new technological landscape and to help encourage the mathematical community to express itself in it as cogently as possible.

As I lay out this mind-boggling new scenario, I endeavor to be as specific as possible. I give lots of concrete examples and plenty of detailed description. At the risk of overstating the obvious, I leave nothing to the imagination.

Fortunately, information and writing are today vastly more accessible to virtually everyone than in past times. But that puts a heavier burden on the writer to be careful, accurate, and responsible. It certainly gives one pause for thought before picking up the pen.

I continue in this new edition the practice from the first edition of labeling mistakes with the symbol ✠ . I hope that this habitude helps the reader to understand what is going on.

I conclude by noting that I am a nontrivial presence in the mathematical publishing world. I have a considerable relationship with Springer, with Birkhäuser, with Taylor & Francis, with the Mathematical Association of America, and with the American Mathematical Society. I have published a great many books with these publishers and with others as well. There is no doubt that my publishing experiences have influenced what I have to say, and I apologize for that in advance. Because of my background I can claim to be knowledgeable, but I cannot claim to be completely objective.

As always, I thank my editor Sergei Gelfand and my readers and colleagues for their support and their friendly, constructive criticism. Sheldon Axler, Don Babbitt, David Bailey, Harold Boas, John P. D'Angelo, Fausto Di Biase, John Ewing, Jerry Folland, Peter Gilkey, David Hoffman, Dmitri Khavinson, Blake Thornton, and Steve Weintraub have been particularly helpful. When it comes to careful use of language, Randi D. Ruden is always my best teacher. She deserves my profound thanks for her careful reading and cogent criticisms.

Special thanks go to Lynn Apfel and Robert Burckel for an especially detailed reading and many incisive remarks, criticisms, and suggestions.

I look forward to reader feedback on this new edition.

S.G.K.
St. Louis, Missouri

# Preface to the First Edition

The past fifty years have not seen as much emphasis on the quality of mathematical writing as perhaps one would wish. Because of competition for grants and other accolades, we hasten our work into print. An obituary for Hans Heilbronn (1908-1975) asserted that, after he wrote (by hand) a draft of a paper, he would put it on the shelf for one year. Then he would come back to it with fresh eyes, read it critically, and rewrite it. In effect, after a year's time, Heilbronn was reading his own work as though he were unfamiliar with it and had to understand each point from first principles. It is perhaps worth dwelling on this exercise to see what we might learn from it.

There is no feeling quite like that which comes after you have proved a good theorem, or solved a problem that you have worked on for a long time. Driven by the heat of passion, the words burst forth from your pen, the definitions get punched into shape, the proofs are built and bent and patched and shored up, and out goes that preprint to an appreciative audience. The whole paper sparkles—both the correct parts and the incorrect parts. A friend of mine, who solved a problem after working on it to the exclusion of all else for over fifteen years, used to rise up in the middle of the night just to caress his manuscript lovingly.

In circumstances like these, you find it virtually impossible to distance yourself from the material. Everything is emblazoned in your own mind and is crystal clear; you are unable to take the part of the uninitiated reader. You are torn between the desire (expeditiously) to record and validate your ideas, and the desire to communicate and explain.

In today's competitive world, you probably do not feel that you have the luxury of setting a new paper aside for a year. The paper could be scooped; the subject could take a different direction and leave your great advance in the dust; the NSF might cancel your grant; the Dean might not give you a raise; you might not be invited to speak at that big conference coming up.

Now let us look through the other end of the telescope. The harsh reality is this: If you prove the Riemann hypothesis, or the three-dimensional Poincaré conjecture, or Fermat's Last Theorem, then the world is willing to forgive you a lot. It is nice if your paper is well written, for then more people will benefit from it more quickly. But—even if the paper is abysmally written—a handful of experts will be able to slug their way through it, they will teach it to others, perhaps more transparent proofs may come out, textbooks will eventually appear. Science is a process that tends to work itself out.

In fact most of us do not produce work at the high level just described—certainly not consistently so. If your work is not written in a clear fashion, so that the reader may quickly apprehend what the paper is about, what the main results are, and how the arguments proceed, then the reader will likely set it aside before reading much of it. Your work will not have the impact that you had hoped or intended.

I am certainly not writing this book to advocate that you set aside each of your papers for a year, in the fashion of Heilbronn, and then rewrite it. Rather, I am asking you to consider the value of learning to write. Heilbronn had his techniques for sharpening up his prose. Each of us must learn our own.

I know many examples of mathematicians $A$ and $B$, of roughly similar talent, with the property that $A$ has enjoyed much greater success than $B$, and considerably more recognition for his/her ideas, because $A$ wrote his/her work in an appealing and readable fashion and $B$ did not. The $A$s and $B$s that I am thinking about are not at the Fields Medal level; Fields Medalists are exceptional in almost every respect, and clearly do not need my advice. Instead, the examples of which I speak are several notches down from that august level, like most of us.

Even if you accept my thesis—that it is worthwhile to learn to write mathematics well—you may feel that fine writing is not an avocation that you wish to pursue. Fair enough: if you had wanted to become a writer, then probably you would have done so. But I submit that a reasonable alternative might be to spend an hour or two with this book, and perhaps another hour or two considering how its precepts apply to your own writing. The result, I hope, will be that you will be a more effective writer and will derive more enjoyment from the writing process.

As a scholar, or a scientist, you do not make widgets, nor do you grow wheat, nor do you perform brain surgery. In fact, what you do is manipulate ideas and report on the results. Usually this report is in written form. What you write is often important, and can have real impact. Freshman composition teachers at Penn State like to tell their students of the engineers at Three Mile Island, who wrote to the governor of Pennsylvania three times to tell him that a nuclear disaster was in the making at their power plant. Their prose was so garbled that the poor governor could not determine *what* in the world they were talking about. The rest is history.

The very act of writing has, in the last twenty years, taken on a new shape and form. Whereas, years ago, it entailed sharpening a quill and buying a bottle of ink, nowadays most of us do not even own a quill knife. Instead we boot up the computer and create a document in some version of TeX. This being the case, I have decided to devote a (large) portion of this book to *techniques* of effective writing and another (much smaller) portion to the *instruments* of modern writing. This book is intended in large part for the novice mathematician. Fresh from graduate school, such a person must engage in the struggle of figuring out how to survive in the profession. The lucky budding mathematician will have gone to a graduate program that provided experience in technical writing and the use of hardware and software. If not, then perhaps that person is presently in a working environment that makes it easy to learn the technical aspects of writing. But I think that it is useful to have a reference for these matters. I intend, with this book, to provide one.

My credentials for writing this book are simple: I have written about one hundred articles and have written or edited about fifteen books. I have received a certain amount of praise for my work, and even a few prizes; and I have received plenty of criticism. Let me assure you that one of the most important attributes of a good writer is an ability to listen to criticism and to learn from it. Everyone finds it difficult to read criticism without becoming defensive; nobody wants to be excoriated. But even the most negative, uncharitable review can contain useful information. You profit not at all by becoming emotional; but if you can use the criticism to improve your work, then you have trumped the critic.

This book is a rather personal tract, containing personal recommendations that reflect my own tastes. I have reason to believe that many others share these tastes, but not all do. There certainly are treatments of the art of mathematical writing that are more scholarly than this one—I note particularly the book [Hig] of Higham. He has careful discussions of how to select a dictionary or a thesaurus, careful catalogings of British usage versus American usage, a history of mathematical notation, clever exercises for developing skill with English syntax, tutorials on revision, and so forth. Higham's book is a real labor of love, and I recommend it highly. But there is no sense for me to duplicate Higham's efforts. Here I will discuss how to write, why to write, and when to write. However, this is not a scholarly tract, and it is not a text. The book is intended, rather, to be some friendly advice from a colleague. If an Assistant Professor or Instructor were to come to my office and ask for suggestions about writing, then I might reply "Let's go to lunch and talk about it." This book encompasses what I would say over the course of several such meetings.

In this book I shall not give an exhaustive treatment of grammar, nor of *any particular aspect* of writing. When I do go into some considerable detail, it is usually on a topic not given extensive coverage elsewhere. Examples of such topics are **(i)** How to organize a paper, **(ii)** How to organize a book, **(iii)** How to write a letter of recommendation in a tenure case, **(iv)** How to write a referee's report, **(v)** How to write a book review, **(vi)** How to write a talk, **(vii)** How to write a grant proposal, **(viii)** How to write your Vita.

I have adopted the practice of labeling *incorrect* examples of grammar and usage with the symbol ✠. I do this so that examples of what is wrong will not be mistaken for examples of what is right.

I have benefited enormously from many friends and colleagues who were kind enough to read various drafts of the manuscript for this book. Their comments were insightful, and in many cases essential. In some instances they saved me from myself. I would like particularly to mention Lynn S. Apfel, Sheldon Axler, Don Babbitt, Harold Boas, Robert Burckel, Joe Christy, John P. D'Angelo, John Ewing, Gerald B. Folland, Len Gillman, Robert E. Greene, Paul Halmos, David Hoffman, Gary Jensen, Judy Kenney, Donald E. Knuth, Silvio Levy,

Chris Mahan, John McCarthy, Jeff McNeal, Charles Neville, Richard Rochberg, Steven Weintraub, and Guido Weiss. George Piranian generously exercised his editing skills on my manuscript, and to good effect. I thank Randi Ruden for sharing with me her keen sense of language and her sharp wit; she showed no mercy, and spared no pains, in correcting my language and my logic. Josephine S. Krantz provided valuable moral support. I find it a privilege to be part of a community of scholars that is so generous with its ideas. Pat Morgan, Antoinette Schleyer, and Jennifer Sharp of the American Mathematical Society gave freely of their copy editing skills. Barbara Luszczynska, our mathematics librarian, also gave me help in tracking down sources. My work at MSRI was supported in part by NSF grant DMS-9022140.

It would be impossible for me to enumerate, or to thank properly, all the excellent mathematical writers from whose work I have learned. They have set the example, over and over again, and I am merely attempting to explain what they have taught us. Several other authors have addressed themselves to the task of explaining how to write mathematics, or how to execute scientific writing, or simply how to write. Some of their work is listed in the Bibliography. (See also [Hig] for a truly extensive enumeration of the literature.) The present book interprets some of the same issues from my own point of view, and filtered through my own sensibilities. I hope that it is a useful contribution.

S.G.K.
St. Louis, Missouri

# Chapter 1

# The Basics

*Against the disease of writing one must take special precautions, since it is a dangerous and contagious disease.*

Peter Abelard
Letter 8, Abelard to Heloise

*Judge an artist not by the quality of what is framed and hanging on the walls, but by the quality of what's in the wastebasket.*

Anon., quoted by Leslie Lamport

*It matters not how strait the gate,*
*How charged with punishments the scroll,*
*I am the master of my fate;*
*I am the captain of my soul.*

W. E. Henley

*Your manuscript is both good and original; but the part that is good is not original, and the part that is original is not good.*

Samuel Johnson

*In America only the successful writer is important, in France all writers are important, in England no writer is important, and in Australia you have to explain what a writer is.*

Geoffrey Cotterel

*It may be true that people who are merely mathematicians have certain specific shortcomings; however, that is not the fault of mathematics, but is true of every exclusive occupation.*

Carl Friedrich Gauss
letter to H. C. Schumacher [1845]

*In fifty years nobody will have tenure but everyone will have a Ph.D.*

M. V. Wickerhauser

## 1.1   What It Is All About

### 1.1.1   How to Write and Why to Write

In order to write effectively and well, you must have something to say. You must also be *confident* that you have something to say, and that that something is worth saying. Finally, you have to figure out how to say it.

This sounds trite, but it is the single most important fact about writing. In order to write effectively and well, you must also have an audience. And you must *know consciously* who that audience is. Much of the bad writing that exists is performed by the author of a math research paper who thinks that all of his/her readers are Henri Poincaré, or by the author of a textbook who does not seem to realize that his/her readers will be students.

Good writing requires a certain confidence. You must be confident that you have something to say, and that that something is worth saying. But you also must have the confidence to know that "My audience is $X$ and I will write for $X$."

Indeed, many writers of a mathematical paper seem to be writing primarily to convince themselves that their theorem is correct, rather than as an effort to communicate. Such authors seem embarrassed to explain anything, and hide instead behind the details.

Many textbook authors seem embarrassed to speak to the student in language that the student will apprehend. Such authors instead find themselves making excuses and asides to the instructor (who either will not read the book, or will flip through it impatiently and entirely miss the author's efforts).

Imagine penning a poem to your one true love, all the while thinking "What would my English teacher think?" or "What would my pastor think?" or "What would my mother think?" Have the courage of your convictions. Define your audience. Doing so, you should say to yourself, "My audience is $X$ and I will write for $X$." Speak to that person or to those people whom you are genuinely trying to reach. Know what it is that you want to say and then say it, all the while anticipating your specific audience's specific needs.

I note here that I am an American author and I am writing with an American audience in mind. So the opinions that I express, and the rules that I enunciate, are American rules. British rules are often different. And no rule of grammar or syntax is etched in stone. These are ever-changing. It is what makes life interesting.

### 1.1.2   The Research Paper

For at least some mathematicians, the most important writing is the writing of a research paper. You have proved a nice theorem, perhaps a great theorem. You certainly have something to say. You also know exactly who your audience is: other research mathematicians who are interested in your field of study. Thus two of the biggest problems for a writer are already solved. The issue that remains is how to put it on paper. Remember that, as much as you might admire your own results, if you pen a love letter to yourself, then it will have both the good features and the bad features of such a screed: it will exhibit both passion and fervor, but it will tend to exclude the rest of the world. What do these remarks mean in practice? In particular, they mean that as you write you must think of your reader—not yourself. Narcissistic writing is precisely that—narcissism is suitable in some contexts, but not in a research mathematics context. As a mathematics writer, you must place yourself at the service of your readers. You must consider their convenience and understanding, not just your own.

In the Sputnik era, in the late 1950s and early 1960s, when mathematics departments and journals were growing explosively and everyone was in a rush to publish, it was common to begin a paper by writing "Notation is as in my last paper." Today, by contrast, there are truly gifted mathematicians who write papers that look like a letter home to Mom: they just start to write, occasionally starting a new paragraph when the text spills over onto a new page, never formally stating a theorem or even a definition, never coming to any particular point. The contents may be divine, but busy readers will likely lack all patience to discover and understand them.

These last are not the sorts of papers that you would want to read, so why torment your readers by writing papers like this? Much of the remainder of this book will discuss ways to write your work so that people *will* want to read it, and will enjoy it when they do so.

## 1.2   Who Is My Audience?

### 1.2.1   Identifying Your Audience

If you are writing a diary, then it may be safe to say that your readership is just yourself. Truthfully, even this may not be accurate. For you may have it in the back of your mind that (like Anne Frank's or Samuel Pepys's diaries) this piece of writing is something for the ages. If you are writing a letter home to Mom, then your audience is Mom and, on a good day, perhaps Pop. If you are writing a calculus exam, then your audience consists of your students, and perhaps some of your colleagues (or your Chair, if the Chair is in the habit of reviewing your teaching). If you are writing a tract on handle-body theory, then your audience is probably a well-defined group of fellow mathematicians (most likely topologists). Know your audience!

Keep in mind a specific person—somebody actually in your acquaint- ance—to whom you might be writing. If you are writing to yourself or to Mom, this is easy. If you are instead writing to your peers in handle-body theory, then think of someone in particular—someone to whom you could be explaining your ideas. This technique is more than a facile artifice; it helps you to picture what questions might be asked, or what confusions or objections might arise, or which details you might need to trot out and explain. It enables you to formulate the explanation of an idea, or the clarification of a difficult point.

## 1.3    Writing and Thought

### 1.3.1    Clear Thinking

The ability to think clearly and the ability to write clearly are inextricably linked. If you cannot articulate a thought, formulate an argument, marshal data, assimilate ideas, or organize a thesis, then you will not be an effective writer. By the same token, you can use your writing as a method of developing and honing your thoughts—your hypotheses, your theorems, your proofs. See [Hig] for an insightful discussion of this concept.

We all know that one way to work out our insights and organize our ideas is to engage in an animated discussion with someone whom we respect. But you can instead, à la Descartes, have that interchange with yourself. And a useful way to do so is by writing. When I want to develop my ideas on some topic—teaching reform, or the funding of mathematics, or the directions that future research in several complex variables ought to take, or my new ideas about domains with noncompact automorphism group—I often find it useful to write a little essay on the subject. For writing forces me to express my ideas clearly and in the proper order, to fill in logical gaps, to differentiate hypotheses from blind assumptions from conclusions, and to make my point forcefully and clearly. Sometimes I show the resulting essay to friends and colleagues, and sometimes, with many edits and revisions, I try to publish it. But, just as often, I file it away on my hard disk and forget about it until I have future need to refer to it.

### 1.3.2    Research Mathematics

The writing of research-level mathematics is a more formal process than that described in the last paragraph, but it can begin in the same way. You can begin your exposition with a little essay that organizes your ideas, sets your course, and defines your goals. This should probably be a draft that only you will see. You can judge this first draft, and indeed subsequent drafts, as any critical reader would. You may, and probably will, find that that "obvious lemma" is not so obvious after all. You may have to put in some extra work to make things come out right. You may, along the way, discover extraneous content, ideas that may have momentary interest but are not really germane to your main thesis.

When you write up your latest ideas for dissemination and publication, then you must finally face the music. That critical lemma must now be treated; the case that you did not really want to consider must be dispatched. The ideas must be put in logical order and the chain of reasoning forged and fixed. It can be a real pleasure to craft your latest burst of creativity into a compelling flood of logic and calculation. In any event, this skill is one that you are obliged to master if you wish to see your work in print, and read by other people, and understood and appreciated.

### 1.3.3   Various Drafts

You might think in terms of writing a sequence of drafts of whatever project you are now working on:

- The first draft just puts the basic ideas down on paper. Be as accurate as you can, but do not get bogged down at this early stage.

- The second draft considers organization, accuracy, flow of ideas, logic, and potential audience reaction.

- The third draft is the proofreading stage. You should focus on syntax, spelling, diction, sound, and grammar.

- By the fourth draft you should have a fairly polished piece in your hands. Read it aloud, listening for meaning and coherence.

- By the fifth draft you can assume that you have said what you want to say. Now check for spelling (a spell-checker can be useful here), typos, grammatical slips, and the like. Make sure that all the pages are formatted properly. Make sure that all the equations and displayed mathematics look as they should.

As a writer, you in some sense resemble the budding tennis pro who is practicing his/her backhand or the future concert pianist perfecting his/her scales. Little by little, the technical exercise of writing becomes less dreary and more pleasurable. And certainly more constructive and effective.

### 1.3.4   Writing as an Enabling Activity

Once you apprehend the principles just enunciated, writing ceases to be a dreary chore and instead turns into a constructive activity. It becomes a new challenge that you can aim to perfect—like your tennis backhand or your piano playing. If you are the sort of person who sits in front of the computer screen befuddled, frustrated, or even angry, and thinks "I know just what my thoughts are, but I cannot figure out how to say them," then something is wrong. Perhaps you are impeding your mathematical creativity by assuming that your first draft is really your final draft (sadly, many if not most of us are guilty of this sin). Yet the first draft is almost inevitably thoughts-in-a-jumble, not as well or thoroughly reasoned out as you believe they are. Clarify your insights and put them in order—certainly the way that you order your ideas reflects critically on your thinking skills. Adopting the advice given here will not only make you a better writer but also a stronger thinker and a better mathematician.

Writing should *enable* you to express your thoughts, not hinder you. I hope that reading this book will help you to write, indeed will enable you to write, both effectively and well.

## 1.4   Say What You Mean, Mean What You Say

### 1.4.1   Obscure Language

You have likely often heard, or perhaps uttered, a sentence like

> As a valued customer of XYZ Co., your call is very important to us.
> ✠

Or perhaps

> To assist you better, please select one of the following from our menu.
> ✠

What is wrong with these sentences? The first suggests that "your call" is a valued customer. Clearly that concept is not the intention here. A more accurately formulated sentence would be

> You are a valued customer of XYZ Co., and your call is very important to us.

Or perhaps

> Because you are a valued customer of XYZ Co., your call is very important to us.

In the second misspoken example, the phrase "To assist you better" lacks a subject; it is clearly intended to modify an invisible "we"; therefore a stronger construction is

> So that we may better assist you, please select an item from our menu . . . .

or perhaps

> We can assist you more efficiently if you will make a selection from the following menu.

What is my point here? Am I just pompously nit-picking? Assuredly not. Mathematics cannot tolerate imprecision. And while the nature of mathematical *notation* tends to rule out imprecision, the *words* that connect our formulae can *lead to* imprecision. If you formulate your thoughts inaccurately, then your point may well be lost. Here are a few more examples of sentences that do not convey what their authors intended:

> Having spoken at hundreds of universities, the brontosaurus was a large green lizard.          ✠

(Amazingly, this sentence is a slight variant of one that was uttered by a distinguished scholar who is world famous for his careful use of prose.)

> As in our food, we strive to be creative with keeping the highest quality in mind, we have in our wine selections also. ✠

(This sentence was taken from the menu of a rather good St. Louis restaurant.)

> To serve you better, please form a line. ✠

(How many times have you heard this at your local retailer's, or at the bank?)

The message here is a simple one: Make sure that your subject matches your verb. Make sure that your referents actually refer to the intended person or thing. Make sure that your participles do not dangle. Make sure that your clauses cohere. *Read each sentence aloud*. Does each one make sense? Would you *say this in a conversation?* Would you understand it if someone else said it?

### 1.4.2 Linear Versus Nonlinear

In life, we receive many different streams of ideas simultaneously, and we parallel-process them in that greatest of all CPUs—the human brain. We absorb and process information and knowledge in a nonlinear fashion. But mathematical writing must strive for linear order. Word $k$ proceeds directly after word $(k-1)$. The distinction between written language as a medium of communication and the information flow that we commonly experience helter-skelter is one of the barriers between you and good writing. I myself struggle with this matter constantly. Indeed, as you read this book (which purports to tell you how to write), you will see passages in which I say "now I will digress for a moment" or "here is an aside." (In other places I put sentences in parentheses or brackets; or I use a footnote.) These are junctures at which I could not fit the material being discussed into strictly logical order.

But I can guarantee that you will have to learn to wrestle with similar problems in your own writing. One version of writer's block is a congenital inability to address this linear vs. nonlinear problem. In this situation, nothing succeeds like success. I recommend that, next time you encounter such a difficulty, address it head on. Find devices to help you work around the block. Work the writer's block into submission by forcing the words to say what you want to say. After you have defeated this problem a few times (not without a struggle!), then you will be confident that you can handle it in the future.

### 1.4.3 Global Issues

I have discoursed on accurate use of language in the technical sense. Now let me remark on more global issues. As you write, you must think not only about whether your writing is correct and appropriate, but also about where your writing will go and what it will do when it gets there.

And you must think about organization. This is one of many ways that you exhibit respect for your audience. Cogent organization makes your writing compelling, helps the reader to be satisfied with the reading experience, and effects successful communication.

I have already admonished you to know when to start writing. Namely, you begin writing when you have something to say and you know clearly to whom you wish to say it. You also must know when to stop writing. Stop when you have said what you have to say. Say it clearly, say it completely, say it forcefully, say it without leap or lacuna, but then shut up. To prattle on and on is not to convince further.

### 1.4.4   Language as a Weapon

When you are a person of some accomplishment, and some clout, then your writing carries considerable responsibility. Your words may have great effect. You must weigh the words, and weigh their anticipated impact, carefully.

It is possible to cause considerable damage with what you write. You can have a considerable, and not very constructive, effect on people's lives. You can change the course of events.

The lesson here is to take writing seriously. Think about how people will read your words, and how they will react to your words. If you are writing something delicate, whose ultimate impact is difficult to predict, then take the time to give the matter careful thought. It is only the sensitive and courteous thing to do.

## 1.5   Diction

Diction is concerned with word choice, and how that choice can affect your meaning and your message. In this rather long section we shall treat a number of different aspects of diction in mathematical writing.

### 1.5.1   Careful Use of Words

Use words carefully. A well-trained mathematician is not likely to use the word "continuous" to mean "measurable" nor "pseudoconvex" to mean "one-connected." However, we sometimes lapse into sloppiness when using ordinary prose. Treat your dictionary as a close friend: consult it frequently. Do not use "enervate" to mean "invigorate" nor "fatuous" to mean "overweight" nor "provenance" to denote a geopolitical entity. When I am being underhanded, it is not because I am short of help.

### 1.5.2   Attitude

And never doubt that language is a weapon. "Sticks and stones may break my bones but words will never hurt me" is perhaps the most foolish sentence ever uttered. You can inflict more damage, sometimes permanently, with words than you can with any weapon. You can manipulate more minds, and more people, with words than with any other device.

When a police officer addresses you by

> Sir, may I see your driver's license? Did you notice that red light
> back there?

then he/she is sending out one sort of signal. (Namely, you are clearly a law-abiding citizen and he/she is just doing his/her job by pulling you over and perhaps giving you a ticket.) When instead a cop in the station house says

> OK, Billy. Why don't you spill your guts? You know that those
> other bums aren't going to do a thing to protect you. All they care
> about is saving their own skins. Jacko already confessed to the heist
> and told us that you held the gun, Billy. Now we need to hear it
> from you. Make it easy on yourself, Billy: play ball with us and
> we'll play ball with you.

then he is sending out quite a different sort of signal. (Namely, by using the first name—and not "William," but "Billy"—he is undercutting the addressee's dignity; he is treating the person like a wayward child. Further, the policeman is cutting off the individual from his peers, making him feel as though he is on his own. He is suggesting—albeit vaguely—that he may be willing to cut a deal.)

## 1.5.3   Corruption of Language

I am going to turn now to a brief homily. (I promise that there will be no additional homilies in the book; you may even ignore this one if you wish.) John Locke said that the most effective way to bring down a society is to corrupt its language. You need only look around you to perceive the truth of this statement. When language is corrupted, then people do not communicate effectively. When they do not communicate effectively, then they cannot cooperate. When they cannot cooperate, then the fabric of civilization begins to unravel.

Some of us use the word "bad" to mean "good." We use the phrase "let us keep our neighborhoods safe and clean" to mean "let us segregate our schools and arm every home"; we use the phrase "I am for gun control and freedom of choice" to mean "I'm a liberal and you're not." We say "account executive" when we mean "sales clerk" and "sanitation engineer" when we mean "garbage man." We use the words "interesting" to mean "foolish," "imaginative" to mean "irresponsible," and "naive" to mean "idiotic." These observations are not just idle cocktail party banter. They are in fact indicative of barriers between certain social groups and of the use of language to manipulate and even to coerce.

It is just the same in mathematics. When we use the word "proof" to mean "guesses based on computer printouts" (see [Hor]), when we use "theoretical mathematics" to mean "speculative mathematics" (see [JQ]), when we use the phrase "Charles mathematicians" to belittle the practitioners of traditional and hard-won modes of reasoning that have been developed over many centuries (see [Ati, pp. 193–196]), when we use the phrase "new mathematics" to mean "facile intuition" (see [PS], [Ati, pp. 193–196]), then we are trivializing our subject. These are gross examples, but the same type of corruption occurs in the small when we write our work sloppily or not at all. It is the responsibility of today's scholars to develop, nurture, and record our subject for future generations. Good writing is of course not an end in itself; writing is instead the means for achieving the important goal of communicating and preserving mathematics.

### 1.5.4    What Is in a Title?

Consider these simple examples. Suppose that the Hemingway novel *For Whom the Bell Tolls* were instead entitled *Who the Dingdong Rings For*; or that the Thornton Wilder play *Our Town* were called *My Turf*. Even though the sense of the titles has not been changed appreciably, we see that the alternative titles eschew all the poetry and imagery we sense in the originals. *For Whom the Bell Tolls* evokes powerful emotions; the proffered alternative falls flat. The title *Our Town* suggests one value system, while *My Turf* brings to mind another. One fancies that, if *The Scarlet Letter* had had a less poetic title (how about *Bad Girls Finish Last*), then perhaps Hester Prynne would have garnered only an "$A-$," or maybe even an "Incomplete."

Mathematicians rarely have to wrestle with these poetic questions. But we need to choose names for mathematical objects; we need to formulate definitions. We need to come up with new terminology. We need to describe and to explain. My Ph.D. advisor thought very carefully about his choices of notation and choices of terminology. He figured that his ideas would have considerable influence and lasting value, and he wanted them to work.

As an instance of these ideas, the word "continuous" is a perfect name for a certain type of function; the alternative terminology "nonhypererratic" would be much less useful. The phrase "the point $x$ lies in a relative neighborhood of $P$" conveys a world of meaning in an elegant and memorable fashion. Not by accident has this terminology become universal. You should strive for this type of precision and elegance in your own writing.

William Shakespeare said that "…a rose by any other name would smell as sweet." This statement is true, and an apt observation by Juliet, in the context of the dilemma that faced Romeo and Juliet. But the name of a person, place, or thing can profoundly affect its future. There will never be a great romantic leading man of stage and screen who is named Eggs Benedict and there will never be a Fields Medalist or other eminent mathematician named Turkey Tetrazzini. The name of an object may not change its properties (consult Saul Kripke's New Theory of Reference for more on this thought), but it can definitely affect the way that the object is perceived by the world at large. Bear this notion in mind as you create terminology, formulate definitions, and give titles to your papers and other works.

Have you ever noticed that, when you are reading a menu or listening to an advertisement, it never fails that the food being described contains "fresh creamery butter" and "pure golden honey"? The marketing people never say "this grub contains butter and honey," for there is nothing appealing about the latter statement. But the first two evoke images of delicious food. As mathematicians, we are not in the position of hawking victuals. But we must still effectively convey our message, and the spirit of that message. We want to inform, and also to inspire.

### 1.5.5 Accuracy in Writing

As I have already advised, do not agonize over each word as you write a first draft. Just get the ideas down on the page. But *do* agonize a bit when you are editing and proofreading. A passage that reads

> This is a very important operator, that has very specific properties, culminating in a very significant theorem.           ✠

is fine as a first try, but does not work well in the long run. It overuses the adverb "very." It does not flow particularly smoothly. It makes the writer sound dull witted. Consider instead the following:

> This operator will be significant for our studies. Its spectral properties, together with the fact that it is smoothing of order 1, will lead to our first fundamental theorem.

The second passage differs from the first in that it has *content*. It communicates with some real specificity. It flows nicely, and makes the writer sound as though he/she has something worthwhile to offer.

An amusing piece of advice, taken from [KnLR, p. 102], is never to use "very" unless you would be comfortable using "damn" in its place!

### 1.5.6 Alliteration

A good, though not ironclad, rule of thumb is not to use the same word, nor even the same sound, in two consecutive sentences. Of course you may reuse the word "the," and the nouns that you are discussing will certainly be repeated; but, if possible, do not repeat descriptive words. In addition, do not place words that sound similar in close proximity.

Be especially wary of alliteration. Vice President Spiro Agnew, with the help of speech writer William Safire, earned for himself a certain reputation by using phrases like "pampered prodigies," "pusillanimous pussyfooters," "vicars of vacillation," and "nattering nabobs of negativism." This rhetoric encouraged, in certain circles, mockery of our vice president—certainly not the reaction that you as a serious mathematician might desire. Lyndon Johnson led us into an escalated Vietnam War by deriding "nervous nellies." To be sure, the alliterative device is often suitable for poetry or other creative writing, and even perhaps for political polemics; but it is almost never appropriate for mathematics. When alliteration is absolutely necessary, for correct mathematical terminology, then you should de-emphasize it.

For example

> This semisimple, sesquilinear operator serves to show sometimes that subgroups of $S$ are sequenced.                              ✠

does not sound like mathematics. The typical reader probably will pause, reread the sentence several times, and wonder whether the writer is putting him/her on. Better is

> Observe that this operator is both semisimple and sesquilinear. These properties can lead to the conclusion that if $G$ is a subgroup of $S$ then $G$ is sequenced.

Notice how the division of one sentence into two is used to break up the alliteration, and in the process enhances comprehensibility.

The last two points—not to repeat words or sounds, and to avoid intrusive alliteration—illustrate the principle of "sound and sense." If you read your work aloud as you edit and revise, then you will pick out offending passages quickly and easily. With practice, you also will learn how to repair them. The result will be clearer, more effective writing.

### 1.5.7  "Hence" and Related Words

It is tempting, indeed it is a trap that we all fall into, to overuse a single word that means "hence" or "therefore." An experienced mathematical writer will have a clutch of words (such as "thus," "so," "it follows that," "as a result," "consequently," and so on) to use instead. A paragraph in which every sentence begins with "hence," or with "therefore," can be uncomfortable to read. Have alternatives at your fingertips.

There are a number of words that are overused by everyone. These include "very," "nice," "get,", and so forth. Use one of your proofreadings to look for words like this and eliminate them.

### 1.5.8  Overused and Unnecessary Notation

In general, you should avoid introducing unnecessary notation. Mary Ellen Rudin's famous statement

> Let $X$ be a set. Call it $Y$.

is funny because it is so ludicrous. But this example is not far from the way we write when we are seduced by notation. Consider, for example

> Let $X$ be a compact, metric subspace of the space $Y$. If $f$ is a continuous, $\mathbb{R}$-valued function on $X$, then it assumes both a maximum and a minimum value.                              ✠

This passage suffers from giving names to the metric space, its superspace, the function, and the target space, and then never using any of them. Slightly better is

> Let $X$ be a compact metric space. If $f$ is a continuous, real-valued function on $X$, then $f$ assumes both a maximum and a minimum value.

Better still is

> A continuous, real-valued function on a compact metric space assumes both a maximum value and a minimum value.

The last version of the statement uses no notation, yet conveys the message both succinctly and clearly.

Paul Halmos [Ste] asserts that mathematics should be written so that it reads like a conversation between two mathematicians who are on a walk in the woods. The implementation of this advice may require some effort. If what you have in mind is a huge commutative diagram, or the determinant of a big matrix whose entries are all functions, then you will likely be unsuccessful in conveying your thoughts orally. You must think in terms of how you, or another reasonable person, would *understand* such a complicated object. Of course such understanding is achieved in bits and pieces, and it is achieved conceptually. Such is how you will communicate your ideas during a walk in the woods.

One corollary of the "walk in the woods" approach to writing is that you should write for a reader who is not necessarily sitting in a library, with all the necessary references at his/her fingertips. To be sure, almost any reader will have to look up a few things. But if the reader must race to the stacks, or boot up the computer and do a `Google` search, at every other sentence, then you are making the job too hard. Your paper is far too difficult to follow. Supply the necessary detail and the proper heuristic so that, even if the reader is not sure of a notion, he/she will be able temporarily to suspend his/her disbelief and move on.

As a parting thought, I shall say this: We want our use of notation to be precise and accurate, but we do not want it to be burdensome. While an overwhelming number of superscripts and subscripts may give an extraordinarily accurate description of the space you are discussing, it is no help if it is too tedious and impenetrable. Use Occam's razor, and only indicate the parameters that are really needed.

### 1.5.9  Effective Use of Notation

An aspect of writing peculiar to mathematics is the use of notation. Without good notation, many mathematical ideas would be difficult to express. Indeed, the development of mathematics in the Middle Ages and the early Renaissance was hobbled by a lack of notation. With good notation, our writing has the potential to be forceful and direct.

But you need to be alert to pitfalls. A common misuse of notation is to put it at the beginning of a sentence or a clause. For example,

> Let $f$ be a function. $f$ is said to be *semicontinuous* if ...                ✠

and

> For all points $x$, $x \in S$.                    ✠

Even in these two simple examples you can begin to apprehend the problem: the eye balks at a sentence or clause begun with a symbol. The practice is unsightly, and you find yourself rereading the passage a couple of times in order to discern the correct sense. Much better is:

> A function $f$ is said to be *semicontinuous* if ...

and

> We see that $x \in S$ for all points $x$.

Observe that both of these revisions are easily comprehended the first time through. This is one of the goals of good writing.

In the same spirit, it is very common for mathematical writers to confuse $f$ and $f(x)$. What is correct is that $f$ is the *name* of a function while $f(x)$ is the *value* of that function at the point $x$. So, for instance, you should not say, "The function $f(x)$ has these properties ...." Correct is "The function $f$ has these properties ...."

Mathematical notation is often so elegant and compelling that we are tempted to overuse, or misuse, it. You must be diligent to catch even small or seemingly innocuous errors in notation, for these can alter your meaning and confuse the reader. For example, the notation in the sentence "If $x > 0$, then $x^2 > 0$" is no hindrance, is easy to read, and tends to make the sentence short and sweet (nonetheless, there are those who would tender cogent arguments for "If a number is positive, then so is its square."). By contrast, the phrase

> Every real, nonsquare $x < 0$ ...                    ✠

is objectionable. The reason is that it is not clear, on a first reading, what is meant. Are you saying that "Every real, nonsquare $x$ is negative" or are you saying "Every real, nonsquare $x$ less than zero has the additional property ...." By strictest rules, the notation $<$ is a *binary connective*. The notation is designed for expressing the thought $A < B$. If such is not the exact phrase that fits into your sentence, then you had best not use this notation.

When you are planning a paper, or a book, you should try to plan your notation in advance. You want to be consistent throughout the work in question. To be sure, we have all seen works that, in Section 9, say "For convenience we now change notation." All of a sudden, the author stops using the letter $H$ to denote a subgroup and instead begins to use $H$ to denote a biholomorphic mapping. Amazingly, this abrupt device actually works much of the time—at least with professional mathematicians. But you should avoid it. If you can, use the same notation for a domain in Section 10 (or Chapter 10) of your work that you used in Section 1 (or Chapter 1). Try to avoid local contradictions—like suddenly shifting your free variable from $x$ to $y$. Try not to use the same character for two different purposes.[1]

This last stipulation is admittedly not always easy to follow. Many of us commonly use $i$ for the index of a sequence or series: $\{a_i\}$ or $\sum_{i=1}^{\infty} a_i$. No problem so far, but suppose that you are a complex analyst, and also use $i$ to denote a square root of $-1$. And now suppose that this last $i$ occurs in some of your sequences and series. You can see the difficulties that would arise. It is best to use $j$ or $k$ as the index of your sequence or series. A little planning can help with this problem, though in the end it may involve a great deal of tedious work to weed out all notational ambiguities.

Many a budding mathematician is seduced by mathematical notation. There was a stage in my education when I thought that all of mathematics should be written without words. I wrote long, convoluted streams of $\forall$, $\exists$, $\ni:$, $\Rightarrow$, $\equiv$, and so forth. This style would have served me well had I been invited to coauthor a new edition of *Principia Mathematica* (see [WR]). In modern mathematics, however, you should endeavor to use English—and to minimize the use of cumbersome notation. Why burden the reader with

$$\forall x, x \geq 0 \Rightarrow \exists y, y^2 = x \qquad \maltese$$

when you can instead say

> Every nonnegative real number has a square root.

### 1.5.10  Words as Objects

Sometimes you need to write a sentence that treats a word as an object. Here is an example:

> We call $\Gamma$ the *fundamental solution* for the partial differential operator $L$. We use the definite article "the" because, suitably normalized, there is only one fundamental solution.

I have oversimplified the mathematics here to make a typographical point. First, when you define a term (for the first time), you should italicize the word or phrase being defined. Second, when you refer to a word (in this case "the") as the object of discussion, then put that word in quotation marks. For a variety of psychological reasons, writers often do not follow this rule. It is helpful to recall W. V. O. Quine's admonition:

> "Boston" has six letters. However Boston has 6 million people and no letters.

---

[1] When André Weil was writing his book *Basic Number Theory* [Wei], he strove mightily to follow this advice. He used up all the Roman letters, all the Greek letters, all the fraktur letters, all the script letters, all the Hebrew letters, and all the other commonly used characters that are seen in mathematics. He ended up resorting to Japanese characters.

Word order can have a serious, if subtle, effect on the meaning (or at least the nuance) of a sentence. Consider the following examples:

Yellow is the color of my true love's hair.

My true love's hair has the color yellow.

The yellow hair of my true love is lovely.

These each say something different. All three sentences have the same *meaning*, but decidedly different emphases. To wit, the first of these sentences stresses the color yellow; the second emphasizes the *blonde's* hair; and the third stresses the yellow hair itself. These differences may seem to be relatively minor, but in mathematics they can make a big difference and change the reader's perception of what you are trying to say. (As an exercise, insert the word "only" into all possible positions in the sentence

I helped Carl prove quadratic reciprocity last week.

and watch the meaning change.)

In mathematics, word order can seriously alter the meaning of a sentence, with the result that the sentence is not immediately understood—if at all. When you proofread your own work, you tend to supply meaning not actually present in the writing; the result is that you can easily miss obscurity imposed by word order. Reading your work aloud can help cut through the problem.

## 1.5.11   Singular Versus Plural

Whenever possible, use singular constructions rather than plural. Consider the sentence

Domains with noncompact automorphism groups have orbit accumulation points in their boundaries.     ✠

First, such a construction does not communicate well: should it be "groups" or "group"? More importantly, do all the domains share the same automorphism group, or does each have its own? Does each domain have several orbit accumulation points, or just one? Clearer is the sentence

A domain with noncompact automorphism group has an orbit accumulation point in its boundary.

## 1.5.12   Big Words and Pretension

Avoid the use of big words when small ones will do. Do not say "peregrinate" for "walk," nor "omphaloskepsis" for "thought," nor "floccinaucinihilipilificate" for "trivialize" unless the longer word conveys some important nuance that the shorter word does not. The urge to so bloviate should be resisted. To indulge in hippopotomonstrosesquipedalian tergiversation is not to show your erudition;

rather, it is to be superficial. Also remember that many of your readers will be foreign born, not native English speakers. Make some effort to write simple, straightforward English that they will easily apprehend. Save your high-flown rodomontade for ceremonial occasions.

Likewise—and I have said this elsewhere in the book—stick to simple sentence structures. Even the subjunctive mood can lead to confusion when it is used in mathematical writing. An example of the subjunctive is

He acted as though he were in a daze.

Notice the choice of verb here. Usually mathematics is formulated in simple, declarative sentences. We do not often lapse into the subjunctive. But when we do so we should exercise special care that the meaning is clear. Generally speaking, let the mathematics speak for itself; do not try to dress it up with fancy language.

### 1.5.13   Foreign Words and Phrases

You can have some fun peppering your prose with *bon vivant* and *Gemütlichkeit* and *ad hominem* and *samizdat*, but the careless use of foreign words and phrases does not add anything to most writing. And it will confuse many readers. Use foreign phrases sparingly. If you do use them, typeset them in italics. (An exception should be made for foreign words like "pâté" and "etc." (short for *et cetera*), which have become standard parts of the English language and should be set in Roman.) The books [Hig], [SG], [Swa] give more detailed treatments of this topic.

Good mathematics is difficult. Do not let your writing be a device for making it more so. Use simple, declarative sentences—short ones. Use short paragraphs, each with a simple point. To understand my meaning, put yourself in the position of the reader. You are slugging your way through a tough paper. You come to the proof of the main theorem. After killing yourself for a couple of hours, you finally come to the crux of the argument. And it is a single, dense paragraph spanning two pages. Such a daunting prospect is truly depressing. You do not want to abuse your readers in this fashion. Break up the ideas into palatable bites. Do *not* inject stumbling blocks like foreign or unfamiliar words.

### 1.5.14   Flippancy and Faddish Prose

And now a note on flippancy. A friend of mine once wrote a truly elegant—and important—book that included the phrase "the reader should review enough functional analysis so that he does not barf [*sic*] at the sight of a Banach or Frechet space." At the reviewer's insistence, the phrase was toned down before publication. Another friend published a book with the phrase "we leave the details of this proof for the mentally infirmed." I would advise against this sort of sarcasm. This suggestion is not simply a nod to propriety. You want to be proud of your work. Remember that your thesis advisor and the authorities in the field are likely to look at it. Such puerile prose is not what you want them to see. Most likely, ten years hence, you will wish fervently that you had not included such phrases. Anyone who continues to grow intellectually will look on his/her work of ten years ago with some disdain. But there is no percentage in adding embarrassment to the mix.

Suit your tone, and your choice of words, to the subject at hand. It might be suitable to use the phrase "He had all the efficiency and dexterity of a ruptured snail" to describe a clumsy waiter; this is probably not appropriate language for describing your thesis advisor.

Finally, stay away from faddish prose. If you say "fraternally affiliated, ethically challenged young male" to mean "gang member" or "peregrinating, fashion-challenged, pulchritudinally advanced hostess" to mean "prostitute," then you may be politically correct today but you will be strictly out to lunch tomorrow. Today, many a writer or speaker wants to work the word "dis" (gang talk for "disrespect") or "flame" (yuppie talk for "disrespect") into his/her prose. This practice is a mistake, because in ten years the words will have no meaning.

Mathematical writing is serious writing. You do not want to be flippant, you do not want to use faddish language, you do not want to crack jokes, you do not want to be careless. You must show respect for your audience and respect for yourself.

By the same token, avoid old-fashioned modes of expression. In 1827 it was appropriate for a physician to diagnose a patient with "falling crud and palpitation of the pluck"; in 1930 it was fashionable for a woman to complain of "the fantods." Today these phrases are meaningless. It might exhibit devotion to Fermat to use "adæquibantur" instead of "=" (as did he), but such a practice would lead to boundless confusion today.

Some American writers think that it is tony to pepper their writing with British English. They use "humour" for "humor," "lorry" for "truck," and "spanner" for "wrench." Such language is out of place, and can only lead to obfuscation. It would be just as foolish for an American cookbook to give recipes for spotted dick, bubble-and-squeak, and stodge. Nobody would know what the author was talking about. Use your dictionary to check that you are using the appropriate American words and spellings.

For the same reasons I advise against using contractions, abbreviations, or slang—at least in formal writing. Even acronyms (abbreviations created by using the first letter of each word—such as AMS for American Mathematical Society) are dangerous (see Section 1.12); use them with caution. We write because we want our thoughts to last, and to be comprehensible both now and in years hence. Do not let language stand in the way of that goal.

### 1.5.15  Pronouns

When I was a child, I once asked a mathematician why mathematics was usually written in the first person plural: "We now prove this"; "Our next task is thus"; "We conclude our story as follows." His rejoinder was "This is so that the reader will think that there are a lot of you."

More seriously, when you are writing up mathematics, then you must make a choice about your expository voice. You can say "I will now prove Lemma 5" or "We will now prove Lemma 5" or "One may now turn one's attention to Lemma 5." Which is correct?

As with many choices in writing, this one involves a degree of subjectivity. Every usage varies according to context. The first option is rarely chosen. Most people consider it pompous and slightly disrespectful. The only instance where I find the first person singular to be a comfortable choice is the following: sometimes at the end of a paper one says "At this time I do not know how to prove Conjecture A." The first person singular is appropriate for this particular statement because in fact the writer is imparting to the reader some specific information about his/her own lack of knowledge. Writing in this fashion, I envision myself speaking to you somewhat informally, as a teacher does to his/her students. It would be misleading, and a trifle affected, to say "At this time *one* does not know . . . ." Likewise for "At this time *we* do not know . . . ." The writer could perhaps say, "At this time it is not known whether . . . ." This last choice could be misleading, however, as it suggests that the writer's ignorance is shared by the world at large.

In this book I often use "you," but such usage would be out of place in a formal mathematics paper. As usual, one has to consider the context and the audience.

The first person plural, or "we," is generally your best choice. Unlike the first person singular, or "I," which sounds elitist and irritating, "we" stresses the participatory nature of the enterprise, and encourages the readers to push on. Moreover, since "we" is what people are accustomed to hearing, it is less prone than one of the other choices to jar their ears, or to distract them. The use of third person singular, or "one," often leaves both writer and reader struggling with awkward sentence structures. If you endeavor to write in that mode, then you will probably find yourself soon breathing a sigh of relief as you abandon it.

With a little craftsmanship, you can avoid entirely the use of the first person in your writing. Rather than say "We now turn to the proof of Lemma 4," instead say "Next is the proof of Lemma 4" or perhaps "The next task is the proof of Lemma 4." Rather than say "We see that the proof is complete," say "The proof is now complete" or "This completes the proof." The book [Dup, Ch. 2] has a sensible and compelling discussion of the question of "We" vs. "I" vs. "one."

Your subject, your purpose, and your audience will usually point you towards which of the words "I," "we," or "one"—or perhaps none of these—you wish to use. I am offering "we" as the default. But the sense of what you are writing may dictate another choice.

## 1.5.16 The Role of English

And now a coda on the role of English in mathematical writing. More and more, English is becoming the language of choice in mathematics. Therefore those of us who are native speakers set the standard for those who are not. We should exercise a bit of care. I have a good friend, also an excellent mathematician, who is widely admired; his fans like to emulate him. He is fond of saying (informally)

> What you need here is to cook up a function $f$ such that . . . .

Mathematicians of foreign extraction, who have been hearing him make this statement for years, have now developed the habit of saying

> Take a function $f$. Now cook it for a while . . . .                    ✠

It is a bit like having your children emulate (poorly) all your bad habits. A word to the wise should suffice.

## 1.5.17  Acronyms

Do not use acronyms, abbreviations, or jargon unless you are dead certain that your audience knows these shortcuts. Speaking of an ICBM, the NAFTA treaty, ARVN, and MIRV is fine for those well read in the current events of the past twenty-five years—and who have an excellent memory to boot. But most of us need to be reminded of the meanings of these acronyms. The best custom is to define the acronym parenthetically the first time it is used in a piece of writing. For example,

> The Strategic Arms Limitation Talks (SALT) were progressing poorly, so we broke for lunch. A few hours later, we resumed our efforts with SALT.

I have served on many AMS (American Mathematical Society) committees, and am somewhat horrified at the extent to which I have become inured to certain acronyms. How many of these do you know: CPUB, COPROF, JSTOR, LRPC, ECBT, COPE? I am conversant with them all, and none has done me a bit of good. In practice, you may not even safely assume that your reader knows what the AMS is—what if he/she is Turkish?

We must keep in mind that there are certain acronyms that most all mathematicians will know (AMS, NSF, NSA, for example) but nonmathematicians probably will not. As always, keep your audience in mind when you write.

## 1.5.18  Jargon

I was once at a meeting to discuss the writing of a new grant proposal—to apply for renewal of funds from a generous source which, we hoped, would be inclined to give again. One of the PIs ("PI" denotes "Principal Investigator") said, in all seriousness, "I think that we are going to need more blue sky in this proposal if we want to generate more bottom line." Of course his meaning was "We must endeavor to paint an enlarged picture of long-term goals and anticipated achievements if we want to increase the size of this grant." The first mode of expression might be appropriate among venture capitalists, who are inured to such language. It is probably inappropriate among academics.

# 1.6 Proofreading, Reading for Sound, Reading for Sense

## 1.6.1 Proofreading

Proofreading is an essential part of the writing process. And it is not a trivial one. (You do not simply write the words and then quickly scan them to be sure that there are no gross errors.) Paul Halmos [Hig] said that he never published a word before he had read it six times. Not all of us are that careful, but the spirit of his practice is correct:

- One proofreading should be for *mathematical accuracy*. Are the theorems correctly stated? Do the proofs cohere? Are the definitions on point?

- One proofreading should be for *organization* and for *logic*.

- One proofreading should be for *sense*, and for the meaning, flow, and integrity of the ideas.

- One proofreading should be to check *spelling* and simple *syntax* errors (software can help with the former, and even with the latter—see Section 6.4).

- One proofreading should be for *sound*.

- One proofreading should be for overall coherence. Does the piece make sense? Does it convince?

It is a good idea, after proofreading your work several times, to put it away for 48 hours or more. Go for a run. Go see a play. Take your spouse out to dinner. Read a good book. Just do something to get your mind going on another track. The point is that you can get so absorbed in your work that you do not see it objectively anymore. You cannot effectively detect errors. You are more like a rubber stamp than a critic. Taking some time away can resharpen your focus and make you a more effective proofreader and critic.

The great English stage actor Laurence Olivier used to rehearse Shakespeare by striding across the countryside and delivering his lines to herds of bewildered cattle. Understandably, you may be disinclined to emulate this practice when developing your next paper on $p$-adic $L$ functions—especially if you live in Brooklyn. However, note this: all the best writers whom I know read their work aloud to themselves. Reading your words aloud *forces* you to slow down, to hear each word and each sentence precisely, to better understand what you have written, and to deliver it as a coherent whole. If you have never tried this technique, then your first experience with it will be a revelation. You will find that you quickly develop a new sensitivity for sound and sense in your writing. You will develop an "ear." You will learn instinctively what works and what does not.

### 1.6.2   Writing with Good Sense

Consider the statement

>    The conjecture of Gauss (1830) is false.        ✠

Contrast this rather bald assertion with

> The lemmas of Euler (1766) and the example of Abel (1827) led
> Gauss to conjecture (1830) that all semistable curves are modu-
> lar. The conjecture was widely believed, and more than fifty papers
> were written by Jacobi, Dirichlet, and Galois in support of it. To
> everyone's surprise and dismay, a counterexample was produced by
> Frobenius in 1902. This counterexample opened many doors.

This second passage puts the entire matter in context, tells the reader who
worked on the conjecture and why, and also how the matter was finally resolved.
Although written mathematics is traditionally terse, at least consider in your
own writing the advantage of telling the reader what is going on.

You will become accustomed to what mathematical writing should look like
by doing a lot of mathematical reading. But you really have to *think* about what
you are seeing in these books and papers. What is the form of the sentences?
What is the form of the paragraphs? What words are most commonly used?
What phraseology is most commonly used? How long is a typical sentence?
How long is a typical paragraph?

A math book does not read like a mystery novel, nor like a restaurant menu,
nor like a religious tract. What makes it different? It is *not* just the notation.
It is the *turns of phrase*, it is the *form* of the writing, it is the *organization*. It
is the *logic*. Mathematical writing is forceful and focused. It is *not* tentative. It
says what needs to be said directly and plainly with simple, incisive sentences.
It proceeds step-by-step. It uses argumentation skillfully and accurately. It is
scholarly and compelling.

## 1.7   Compound Sentences, Passive Voice

### 1.7.1   Overly Complex Sentences

It would be splendid if we could all write with the artistry of Flaubert, the
elegance of Shakespeare, and the wisdom of Goethe. In mathematical writing,
however, such an abundance of talent is neither necessary nor called for. In
developing an intuitionistic ethics ([Moo]), for example, one presents the ideas
as part of a ritualistic dance: there is a certain intellectual pageantry that comes
with the territory. In mathematics, by contrast, what is needed is a clear and
orderly presentation of the ideas.

Mathematics is already, by its nature, logically complex and subtle. The
sentences that link the mathematics are usually most effective when they are
simple, declarative sentences. Compound sentences (two or more independent

clauses joined by a coordinating conjunction such as "and" or independent clauses joined to a dependent clause by a subordinating conjunction such as "although") should be broken up into simple sentences. Avoid run-on sentences at all cost. Here are some examples:

Rather than saying

> As $x$ tends ever closer to $x_0$ then $f(x)$ approaches $f(x_0)$, from which we see that $f$ is continuous and in fact one can use this argument to see that $f$ is uniformly continuous.    ✠

instead say

> As $x \to x_0$ we see that $f(x) \to f(x_0)$. Thus $f$ is continuous. The same reasoning shows that $f$ is uniformly continuous.

Of course mathematical notation allows us to write $\lim_{x \to x_0} f(x) = f(x_0)$ instead of either of these first phrases; this abbreviated presentation will, in many contexts, be more desirable.

Rather than saying

> If $g$ is positive, $f$ is continuous, the domain of $f$ is open, and we further invoke Lemma 2.3.6, then the set of points at which $f \cdot g$ is differentiable is a set of the second category, provided that the space of definition of $f$ is metrizable and separable.    ✠

instead say

> Let $X$ be a separable metric space. Let $f$ be a continuous function that is defined on an open subset of $X$. Suppose that $g$ is any positive function. Using Lemma 2.3.6, we see that the set of points at which $f \cdot g$ is differentiable is of second category.

An alternative formulation, even clearer, is this:

> Let $X$ be a separable metric space. Let $f$ be a continuous function that is defined on an open subset of $U$ of $X$. Suppose that $g$ is any positive function on $U$. Define $S \subset U$ to be the set of points $x$ such that $f \cdot g$ is differentiable at $x$. Then, by Lemma 2.3.6, $S$ is of second category in $U$.

Note the use of the words "suppose" and "define" to break up the monotony. Observe how the formal definition of the set $S$ clarifies the slightly awkward construction in the penultimate version of our statement.

## 1.7.2   Passive Voice

Most authorities believe that writing in the passive voice is less effective than writing in the active voice. To write in the active voice is to identify the agent of the action, and to emphasize that agent acting on the subject (see [Dup] for a powerful discussion of active voice vs. passive voice). In the passive voice, the subject is acted upon. In plain English,

> The dog ate the cat.

is active voice. By contrast,

> The cat was eaten by the dog.

is passive voice.

For a mathematical example, consider

> The manifold $M$ is acted upon by the Lie group $G$ as follows:   ✠

is less direct, and requires more words, than

> The Lie group $G$ acts on the manifold $M$ as follows:

Likewise, the statement

> It follows that the set $Z$ will have no element of the set $Y$ lying in it.         ✠

can be more clearly expressed as

> Therefore no element of $Y$ lies in $Z$.

Even better is

> The sets $Y$ and $Z$ are disjoint.

or

> Therefore $Y \cap Z = \emptyset$.

Notice that the last version of the statement used 1 word, while the first version used 17. Also, a mathematician much more readily apprehends $Y \cap Z = \emptyset$ than he/she does a string of verbiage. Finally, coming up with the succinct fourth formulation required not only restating the proposition, but also thinking about its meaning. The result was plainly worth the effort.

In spite of these examples, and my warnings against passive voice, I must admit that passive voice gives us certain latitude that we do not want to forfeit. If, in the first example, you have reason to stress the role of the manifold $M$ over the Lie group $G$, then you may wish to use passive voice. In the second example, it is unclear how the use of passive voice could add a useful nuance to your thoughts.

# 1.8 Technical Aspects of Writing a Paper

## 1.8.1 Details of Your Draft

Even when your paper is in draft form, your name should be on it. A date is helpful as well. Number the pages. Write on only one side of the paper. Give the paper a working title.[2]

Is all this just too compulsive? No.

First, you must always put your name on your work to identify it as your own. If it contains a good idea, then you do not want someone else to walk off with it. Because you tend to generate so many different drafts and versions of the things that you write, you should date your work. I have even known mathematicians who put a time of day on each draft. (Of course a computer puts a date and time stamp on each *computer file* automatically; here I am discussing hard copy or paper drafts.)

Your academic affiliation should appear—even on the draft. If you are usually at Harvard, then write that down. If instead you are spending the year in Princeton, write that down. The draft could, at some point, be circulated. People need to know where to find you. With this notion in mind, include your current email address.

## 1.8.2 Numbering Systems

Take a few moments to think about the numbering of theorems, definitions, and so forth. This task is important both in writing a paper and in writing a book. Some authors number their theorems from 1 to $n$, their definitions from 1 to $k$, their lemmas from 1 to $p$, their corollaries from 1 to $r$—each item having its own numbering system. Do not laugh: this describes the default system in LATEX. As a reader, I find this method maddening; for the upshot is that I can never find anything. For instance, if I am on the page that contains Lemma 1.6, then that gives me no clue about where to find Theorem 1.5. If, instead, all displayed items are numbered in sequence—Theorem 1.2 followed by Corollary 1.3 followed by Definition 1.4, etc.—then I always know where I am.

Having decided on the logic of your numbering system, you also need to decide how much information you want each number to contain. What does this mean? My favorite numbering system (in writing a book) is to let "⟨⟨*Item*⟩⟩ 3.6.4" denote the fourth enunciated item (theorem or corollary or lemma or definition) in the sixth section of Chapter 3. If there is a labeled, displayed equation in the statement of the ⟨⟨*Item*⟩⟩, then I label it (3.6.4.1). The good feature of this system is that the reader always knows precisely where he/she is, and can find anything easily. The bad feature is that the numbering system is a bit cumbersome. Other authors prefer to number displayed items within each section. Thus, in Section 6 of Chapter 3 the displayed items are numbered simply 1, 2, 3, .... When reference is later made to a theorem, the reference

---

[2]Some mathematicians do their composition directly on the computer. They never pick up a pen or pencil. But the basic principles and admonitions presented here still apply.

is phrased as "by Theorem 4 in Section 6 of Chapter 3" or "by Theorem 4 of Section 3.6." As you can see, this ostensibly simpler numbering system is cumbersome in its own fashion.

The main point is that you want to choose a numbering system that suits your purposes, and then to use it consistently. You want to make your book or paper as easy as possible for your reader to study and navigate. Achieving this end requires that you attend to many small details. Your numbering system is one of the most important of these.

A final point is this: do not number every single thing in your manuscript. This dictum applies whether you are writing a paper or a book. I have seen mathematical writing in which every single paragraph is numbered. Such a device certainly makes navigation easy. But it is cumbersome beyond belief. Likewise, do not number all formulas. You will only be referring to some of them, and the reader knows that. To number all formulas will create confusion in the reader's mind; he/she will no longer be able to discern what is truly important and what is less so.

### 1.8.3   Use of Paper and Ink

As I have already mentioned, when writing your draft (by hand), write on one side of the paper only. If you do not, and if you are writing something fairly technical and complicated (like mathematics), then you can become hopelessly confused when trying to find your place. In addition, you will find that you must frequently set two pages side by side—for the sake of comparing formulas, for instance. This move is easy with a manuscript written on one side, and nearly impossible with one written on two sides.

If you are scrupulous about not wasting paper, and insist on using both sides, then my advice is this: write drafts of your mathematical papers on one side of fresh paper. When that work is typed up and out the door, boldly $X$-out the writing on the front side of each page of your old drafts. Turn the paper over, and use it as scratch paper, or for your laundry list, or perhaps for the first draft of your next paper.

I suggest writing in ink. Pencil can smear, and erasing can tear the page, and it is difficult to read a palimpsest. Also pencil-written material does not photocopy well. Blue pens do not photocopy well either. I always write with a black pen on either white or yellow paper. I write either with a fountain pen or a rolling writer or a fiber-tip pen so that the pen strokes are *dense* and *sharp* and *dark*. I write with a pen that does not skip or blot. If it begins to do either, I immediately discard it and grab a new one.

Of course you cannot erase words that are written with a pen; but you can cross them out, and such a practice is much cleaner. It is easier to read a page written in bold black ink, and which includes some crossed out passages, than to decipher a page of chicken scratch layered over erased smears written with a pencil or written with a pen that is not working properly.

Be sure that your desk is well stocked with paper, pens, Wite-Out®, Post-it notes®, a stapler, staples, a staple remover, cellophane tape, paper clips, manila folders, manila envelopes, scissors, a dictionary, and anything else you may need for writing. Have them all at your fingertips. You do not want to interrupt the precious writing process by running around and looking for something trivial.

Do not write much on each page. I advise writing *large*, and double- or triple-spaced. The reason? First, you want to be able to insert passages, make editorial remarks, make corrections, and so forth. Second, a page full of cramped writing on every line is hard to read. Third, you can more easily rearrange material if there is just a little on each page. For example, if one page contains the statement of the main theorem and nothing else, another contains key definitions and nothing else, and so forth, then you can easily change the location of the main theorem in the body of the paper. If the main theorem is buried in a page with a great deal of other material, then moving it would involve either copying, or photocopying, or cutting with scissors.[3] I read recently of a famous novelist who writes his books on $3'' \times 5''$ index cards. Really! This way he has about one sentence on each card, and it is easy to move them around.

Do not hesitate to use colored pens. For instance, you could be writing text in black ink, making remarks and notes to yourself (like "find this reference" or "fill in this gap") in red ink, and marking unusual characters in green ink. This may sound compulsive, but it makes the editing process much easier.

### 1.8.4 Bibliographic References

A good bibliography is an important component of scholarly work (more on bibliographies can be found in Sections 2.6, 5.5). Suppose that you are writing a paper with a modest number of references (about 25, say), and you are assigning an acronym to each one. For instance, [GH] could refer to the famous book by Griffiths and Harris. When you refer to this work while you are writing, use the acronym. Keep a sheet of notes to remind yourself what each acronym denotes. Do not worry about looking up the detailed bibliographic reference while you are engaged in writing; instead, compartmentalize the procedure. When you are finished writing the paper, you will have a complete, *informal* list of all your references. You can go to `MathSciNet` (Section 7.2) online and find most of your references in an instant. You can also go to your library's catalog online to find locally obtainable references. LaTeX can be a great help in eliminating much of the tedium of assembling and formatting bibliographies. See the discussion in Sections 2.6 and 5.5.

### 1.8.5 The Writing Process

Finally, let me make a few general remarks about the writing process. As you are writing a paper, there will be several junctures at which you feel that you need to look something up: either you cannot remember a theorem, or you have

---

[3]Of course, if you are writing on a computer, then all cutting and pasting and moving of passages is trivial. You can have several windows open at once, and can move from one part of the document to another with ease.

lost a formula, or you need to imitate someone else's proof. My advice is *not* to interrupt yourself while you are writing. Take your red pen and make a note to yourself about what is needed. But *keep writing*. When you are in the mood to write, you should take advantage of the moment and do just that. Interrupting yourself to run to the library, or for any other reason, is a mistake.

Write on a desk free of clutter. It is romantic, to be sure, to watch a film in which the writer labors furiously on a desk awash with papers, books, hamburger bags, ice cream containers, old coffee cups, last week's underwear, and who knows what else. Leave that stuff to the movies. Instead imagine tearing into page 33 of your manuscript and accidentally spilling a week-old cup of coffee and a piece of pepperoni pizza all over your project. Think of the time lost in mopping up the mess, separating the pages, trying to read what you wrote, recopying your pages, and so forth. Enough said.

If you are going to drink coffee or a soda or eat a sandwich while you work, I suggest having the food on a small separate side table. This little inconvenience will force you to be careful, and if you do have an accident then it will not make a mess of your work.

Write in a place where you can concentrate without interruption. Whether you have music going, or a white noise machine playing, or a strobe light flashing is your decision. But if you are going to concentrate on your mathematics, it may take up to an hour to get the wheels turning, to fill your head with all the ideas you need, and to start formulating the necessary assertions. After you have invested the necessary time to tool up, you want to use it effectively. Therefore you do not want to be interrupted. Close the door and unplug the telephone if you must. Victor Hugo used to remove all his clothes and have his servant lock him in a room with nothing but paper and a pen. Moreover, the servant guarded the door so that the great man would not be interrupted by so much as a knock. This method is not very practical, and is perhaps not well suited to modern living, but it is definitely in the right spirit.

## 1.9   More Details of Mathematical Writing

### 1.9.1   Effective Sentence Structure

For the most part, the writing of mathematics is like the writing of English prose. Indeed, it *is* a part of the writing of English. (*Caveat:* I hope that my remarks have some universality, and apply even if you are writing mathematics in Tagalog or Coptic or Lingit.) If you read your work aloud (I advocate this practice in Section 1.5), then you should be reading complete sentences that flow from one to the next, just as they do in good prose.

It is all too easy to write a passage like

Look at this here equation:

$$x^n + y^n = z^n. \quad \maltese$$

If you read your work aloud to yourself—being sensitive both to large and small issues—then you will catch a blunder like this one immediately. Much smoother is the passage

> The equation
> $$x^n + y^n = z^n$$
> tells us that Fermat's Last Theorem is still alive.

Another example of good sentence structure is

> Since
> $$A < B\,,$$
> we know that ....

Notice that the sentence (fragment) reads well aloud: "Since $A$ is less than $B$ we know that ...." The phrase is compact and straightforward and gets its message across directly.

## 1.9.2 Overuse of Commas

Do not overuse commas. I become distressed when I see a sentence like

> We went to the store, to buy some potatoes.          ✠

This sentence requires no punctuation but the period at the end. Slightly more subtle, but still irksome, is

> Now that we have our hypotheses in place, we state our theorem, with the point in mind, that we wish to understand the continuity, of functions in the class $\mathcal{S}$.          ✠

We certainly use a comma to indicate a pause. But the comma indicates a *logical pause*, not a lack of air or lack of good sense. Read the last displayed sentence out loud, with suitable pauses where the commas occur. It sounds like someone huffing and puffing; the pauses have no reason to them. This sentence is not a representative example of the way that we speak, hence it is not indicative of the way that we should write. Much more attractive is

> Our hypotheses are now in place, and we next state our theorem. The point is to understand the continuity properties of functions belonging to the class $\mathcal{S}$.

While there is no universal agreement on the placement of commas, there are logical guidelines that one should follow.

### 1.9.3  Miscellaneous Stipulations

Mathematicians like the word "given." We tend to overuse and misuse it—especially in instances where the word can be discarded entirely. Consider the example "Given a metric space $X$, and a point $p \in X$, we see that ...." More direct is "If $X$ is a metric space and $p \in X$, then ...." We are often tempted to transcribe spoken language and call that written language; such laziness should be defeated. Our misuse of "given" is an example of such sloth.

When you are putting the final polish on a manuscript, look it over for general appearance. In mathematical writing, several consecutive pages of dense prose are not appealing, nor are several consecutive pages of tedious calculation. For ease of reading, the two types of mathematical writing should be interwoven. It requires only a small extra effort to produce a paper or book with comfortable stopping places on every page. The reader needs to take frequent breathers, to survey what he/she has read, to pause and look back. Make it easy for him/her to do so.

While you are thinking about the counterpoint between prose and formulas, think also about the use of displayed math versus in-text math (in TeX (see Section 6.5), the former is set off by double dollar signs $$ while the latter is set off with single dollar signs). Long formulas are usually better displayed, for they are difficult to read when put in text. Of course *important* formulas should be displayed no matter what their length—and provided with numbers or labels if they will be mentioned later. Do not display every single formula, for that will make your paper a cumbersome read. Also do not put every formula in text, as that will make your writing tedious. A little thought will help you to strike a balance, and to use the two formats to good effect.

## 1.10  Essential Rules of Grammar, Syntax, and Usage

### 1.10.1  Introduction

I have intentionally put this discussion of the rules of grammar, syntax, and usage at the end of Chapter 1. The reasons are several. I want, in a gentle way, to de-emphasize them. I do not, however, wish to trivialize them. I am not one of those who says "the battle against 'hopefully' is lost," "the battle for 'which' vs. 'that' is lost," "the battle for 'lay' vs. 'lie' is lost," and so forth. I find such statements facile, and they miss the point that careful writing requires some precision. The argument "You know what I mean; whether I use 'that' or 'which' is incidental" abrogates the fact that accurate writing, and accurate expression of your thoughts, requires accurate use of language. It takes some linguistic skill to recognize that minuscule errors in usage can change or obfuscate your intended meaning.

The intent of this book is that you should learn to write logically and cogently; to say precisely what you mean, using the right words and the right number of words; to eschew obfuscation. You want to develop an ear, so that clear writing becomes natural. To be sure, exact use of the language is not the primary goal of most mathematicians; but it must certainly be a secondary goal, absolutely crucial to the successful dissemination of your ideas.

Fortunately, most of the rules of English usage are succinct and logical. A particularly concise enunciation of the basic rules appears in [SW]. Since I cannot improve on that presentation, I certainly shall not repeat all of its insights. Here I shall mention just a few sticky points that come up frequently in the writing process. I hope that you will find this section and the next to be a useful "quick-and-dirty" reference. With that goal in mind, I have presented the topics in alphabetical order. See also [Chi], [Dup], [Fow], [Fra], [Hig], and [MW] for a more thorough treatment of issues of grammar, syntax, and usage.

Bear in mind, as you read these precepts, that no rule of English grammar is etched in stone. There will certainly be times that a sentence or phrase formed according to the strictest rules will sound just awful. In such an instance, you must override the rules and use your good sense and taste. More will be said about this technique as the book develops.

## 1.10.2 Rules of Grammar and Syntax

Now for some rules:

- **All, Any, Each, Every** In mathematics we commonly formulate statements such as "Show that any continuous function $f$ on the interval $[0, 1]$ has a point $M$ in its domain such that $f(M) \geq f(x)$ for $x \in [0, 1]$." For cognoscenti it is clear that, when we say "any" here, we mean "all." But for others—for students, or for non-native speakers—this slight abuse of language could cause confusion. For example, a student reading this sentence could (perfectly correctly) construe it to mean "Demonstrate that for *some* function $f$ ...." Thus, if this sentence were part of an exercise, the student might answer

  The function $f(x) = -(x - 1/2)^2$ is continuous on $[0, 1]$ and the point $M = 1/2$ satisfies the conclusion. ✠

  The lesson? Avoid using "any" when "all" or "each" or "every" is intended.

  Conversely, even when you are writing for experts you can cause confusion by misusing quantifiers. Avoid using "all" when "every" or "each" is intended. Experts themselves can be confused by far-too-common sentences like

  All continuous functions have a maximum. ✠

  Notice that the sentence suggests that all continuous functions share the *same* maximum. Of course what was intended was

Every continuous function has a maximum.

or, more precisely,

Each continuous function has a maximum.

(Once again we see the advantage, from the point of view of clarity, of the singular over the plural.) As you proofread your work, you must learn to take the part of the reader (who is not *a priori* sure of what is being said) in order to weed out misused quantifiers.

- **Brevity** Endeavor to formulate your thoughts briefly and succinctly. For example, you *could* say

    In point of fact, we devolved upon the decision to solicit opinions, form an enumeration, and produce a tally.                    ✠

Such a sentence sounds mellifluous, sanguine, and high toned. But why not instead say

    We decided to take a vote.

The second sentence says in 6 words what the first said in 19; and it presents the message more clearly and forcefully. Strunk and White [SW] give a thorough and engaging treatment of the topic of brevity, and they speak particularly cogently of eliminating extra or extraneous words. Mathematics is difficult to read under the best of circumstances. Do not make the reader's job even more difficult by weighing down your prose with excess baggage.

I once saw a sign in the elevator of a Washington, D.C., hotel that said

    Do not carry lighted tobacco products in the elevator.              ✠

I can only suppose that some politician created this sign. Why not just say

    No smoking.

Being concise and to-the-point is not simply a pose. It is essential to good writing and effective communication. You do not want to omit important details, but you also want your text to be of a "high information" nature. You should think of your reader as a quite impatient person who will be easily turned off by a wordy, vague, expendable sentence or a paragraph with no useful information. You should ask yourself whether each sentence is worth the space that it occupies. Does it really say anything? Are we better off without it?

- ***British Spelling vs. American spelling***   Many readers of this book will be American, and will be inured to American spelling. But, in an effort to be tony, we are sometimes tempted to write "armour" instead of "armor," "aluminium" instead of "aluminum," and "centre" instead of "center." Please resist. There is no place for British spelling in an American document (and vice versa). It adds nothing, and can only cause confusion.

- ***Comprise* vs. *Compose***   People use the word "comprise" because they think it makes them sound tony. Unfortunately, because most everyone misuses the word, they instead sound uneducated. The correct use of the word "comprise" is

  > The standing committee comprises two women, three men, and a donkey.

  The formula is "*A* comprises *B*." What people often say, or write, instead is

  > The committee is comprised of two women, three men, and a donkey.   ✠

  What *should* have been used in this last instance is "composed," *not* "comprised." Never say "is comprised of."

- ***Contractions***   Do *not* use contractions in formal writing. Thus the words "don't," "can't," "shouldn't," "I'm," "you're," etc., are taboo. Of course you should never write "ain't." You also should avoid abbreviations. Particularly avoid using informal abbreviations like "cuz" for "because," "tho" for "though," and so forth. You will probably never be tempted to work "bar-b-q" into your next paper on para-differential operators, but you might be tempted to use "rite inverse." Please resist.

    More generally, do not use colloquial language. This is confusing for foreigners (and you will have *many* foreign readers) and annoying for native speakers. It is also a good idea to avoid words like "get." Strictly speaking, "get" is not colloquial. But it is an overused word that has too many meanings. There are so many other words that communicate the idea more precisely.

    Occasionally you will find it suitable to use contractions in various kinds of *informal* writing. It can be a way of drawing in your audience, or of warming yourself up to your subject. For example, in the book [Kr2], I intentionally used an occasional contraction in an effort to create a friendly air about the book. By contrast, the present book is a book about writing, and I wish to set a more formal example—so there are no contractions.

- **Denote**   Use the word "denote" carefully. It has a special purpose in mathematics (and in logical positivist philosophy and modal logic); we should take care to preserve its usefulness. Suppose that a certain mathematical symbol $A$ stands for, or represents, the item or set of ideas $B$ (ideally, you should be able to excise any occurrence of $A$ and replace it with $B$ and preserve exactly the intended meaning). Under these circumstances, and *only under these circumstances*, do we say that "$A$ denotes $B$." For example,

  > Let $X$ denote the set of all semisimple homonoids with stable quonset hut.

  If the above discussion seems obvious, then consider the following shade of difference:

  > **(1)** Let $f$ be a continuous function.

  and

  > **(2)** Let $f$ denote a continuous function.          ✠

  The intended meaning of the first sentence here is "let $f$ be *any* continuous function." Thus the first statement is both customary and correct. The second is neither customary *nor* correct. For we use "denote" when we want to say that a certain specific item stands for some other specific item. This is not what we are trying to say here.

  Lack of familiarity with English, or lack of familiarity with the precise meaning of "denote," sometimes leads to dreadful abuses of the word. A common one is "Denote $X$ the set of all left-handed polyglots." I leave it to you to decide whether failing English or failing intellect might be the correct provenance of such a sentence; the lesson for you is not to use "denote" in such a fashion.

  The word "connote," rarely used in mathematical writing, can be (but should not be) confused with "denote." The dictionary teaches us that "$A$ connotes $B$" means that $A$ *suggests* $B$, but not in a logically direct fashion. For example,

  > To a young man, "love" connotes flowers, beautiful music, and happiness.

  is an appropriate use of the word "connote."

- **Enervate**   Often we are lazy, and we use a word according to how it sounds, rather than according to what it actually means. This text offers "enervate" as an instance of this phenomenon. What the word actually *means* is "to lessen the vitality or strength of." But, intuitively, we confuse "enervate" with "energize" and give it essentially the opposite meaning. The lesson is to be careful with words with which you are unfamiliar.

- **He** and **she** It used to be the custom that, if one referred to an abstract person in one's writing, then one used the male pronouns "he" or "him" or "his." These were routinely used in scholarly writing. Now this practice is considered to be politically incorrect. One must treat women the same as men.

  One solution to this problem is to replace "he" with "he/she," "him" with "him/her," and "his" with "his/her." But this practice is a bit clumsy. Another possibility is to replace "he" with "she," replace "him" with "her," and replace "his" with "her." This does not really seem to solve the problem; instead it replaces one conundrum with another. A third possibility, commonly taught at colleges and universities, is to replace the gender-specific pronouns with "they," "them," and "their." This unfortunately results in some rather awkward constructions. A fourth possibility is to preclude all offense by using the words dreamed up by Michael Spivak [SPI]. Spivak replaces "he" and "she" with "e," replaces "him" and "her" with "em," and replaces "his" and "her" with "eir."

  The really best approach, though it requires some extra time and effort, is to phrase your sentences so that they omit pronouns altogether. As an example, instead of saying

  Ask her whether she wants a new computer.

  one could say

  Ask whether a new computer is desired.

- **Hyphen** vs. **en dash** It is common in mathematics, if two mathematicians have proved a result together, to call it something like "the Riemann-Lebesgue lemma." Nowadays this is considered to be inappropriate. The use of the hyphen here may suggest that Riemann and Lebesgue have more than a professional relationship. More grammatically correct in today's climate is to write "the Riemann–Lebesgue lemma." What is the difference? In the second example I used the so-called *en dash* rather than the hyphen. The en dash is a dash which is about the width of the letter "n" in the current font, and it typically indicates spans or differentiation, where it may be considered to replace "and" or "to." For instance, the en dash can be used to denote a range of numbers (as in "pages 324–386"). It is also used in phrases like "the U.S.–Canada border." It carries less emotional baggage and is therefore a better choice to denote a mathematical collaboration.

You may think this discussion ludicrous, but I can tell you that, if you do not conform to the prescription described in the last paragraph, then your copy editor will change all your hyphens to en dashes.

It may be worth mentioning, and Knuth emphasizes this point in the TEXBook, that there are three types of dashes in typesetting. The hyphen, rendered in TEX with a single -, is a punctuation mark used to join words and to separate syllables of a single word. We might say "semi-continuous" or "egg-beater" or "two-thirds majority." And, as we all know, hyphens are used to indicate a broken word at the end of a line of type. The en dash (typically the width of the letter "n" in the current font), typeset in TEX as --, is used to indicate a span or differentiation. The em dash (typically the width of the letter "m" in the current font), typeset in TEX as ---, denotes a break in a sentence or a parenthetical remark. There is also the minus sign—often longer than an em dash—but that is part of mathematics and not of English.

- **If ... Then**  The most important logical syllogism for the mathematician is *modus ponendo ponens*, or "if ... then." If you begin a sentence with the word "If," then do not forget to include the word "then." Consider this example:

     If $x > 4$, $y < 2$, the circle has radius at least 6, the sky is blue,
     the circle can be squared.                   ✠

Which part of this sentence is the hypothesis and which the conclusion? After a few readings you may be able to figure it out. If it were sensible mathematics, then the mathematical meaning would probably give you some clues. But it is clearer to write

     If $x > 4$, $y < 2$, the circle has radius at least 6, and the sky is
     blue, then the circle can be squared.

Following the dictum that shorter sentences are frequently preferable to longer ones, you can express the preceding thought even more succinctly as

     Suppose that $x > 4$, $y < 2$, the circle has radius at least 6, and
     the sky is blue. Then the circle can be squared.

The word "Then" is pivotal to the logical structure here. It acts both as a connective and as a sign post. And never doubt that such linguistic clues are absolutely necessary. For although the reader can (often) figure out what is meant if the word "then" is omitted, he/she should not *have* to do so. Your job as the writer is to perform this task *for* the reader, to make the reader's job easier. You want your audience to concentrate on the beauty of your mathematics, not on the ambiguities of your prose.

Mathematicians have a tendency to want to jam everything into one sentence. However, as the last example illustrates, greater clarity can often be achieved by breaking things up; this device also forces you to think more clearly and to organize your thoughts more effectively.

Mathematicians commonly write "If $f$ is a continuous function, then prove $X$." A moment's thought shows that this is not the intended meaning: the desire to prove $X$ is not contingent on the continuity of $f$. What is intended is "Prove that, if $f$ is a continuous function, then $X$." In other words, the hypothesis about $f$ is part of what needs to be proved.

The phrase "if and only if" is a useful mathematical device. It indicates logical equivalence of the two phrases that it connects. While the phrase is surely used in some other disciplines, it plays a special role in mathematical writing; we should take some care to treat it with deference. Some people choose to write it as "if, and only if,"—with two commas. Such a practice is perfectly kosher, if a little stilted. One unacceptable habit (because it sounds artificial and is difficult to read) is beginning a sentence with this phrase. For instance,

> If and only if $x$ is nonnegative, can we be sure that the real number $x$ has a real square root. ✠

What a painful sentence to read, whether the reading is done aloud or *sotto voce*. Better is

> A real number $x$ has a real square root if and only if $x \geq 0$.

An alternative form, not with universal appeal (but better than beginning a sentence with "if and only if"), is

> Nonnegative real numbers, and only those, have real square roots.

It is a fact that many definitions ought to be formulated using "if and only if." For instance,

> **Definition:** The function $f$ is *continuous* at the point $P$ in its domain if and only if $\lim_{x \to P} f(x) = f(P)$.

But we often, either out of habit or out of laziness, write "if" instead of "if and only if." It is too bad, because this can confuse neophytes and non-English speakers. Strive to be careful about this matter.

Incidentally, the neologism "iff," reputed to have been popularized by Paul Halmos, is a generally accepted abbreviation for "if and only if." It provides a useful bridge between the formality of "if and only if" and the convenience of "if." It is also common to use the symbol $\iff$ for "if and only if."

- **If vs. Whether**  The words "if" and "whether" have different meanings, and are suitable for different contexts. Follow the example of master editor George Piranian:

  Go to the window and see *whether* it is raining; *if* it is raining, then let Fido inside.

- **Infer and Imply**  The words "infer" and "imply" are often confused in everyday usage. It should not be difficult for a mathematician to keep these straight. A set of assumptions can *imply* a conclusion. But one *infers* the conclusion from the assumptions. It is that simple.

- **Its and It's**  Use "it's" only to denote the contraction for "it is." Otherwise use "its." For example "Give the class its exam" and "A place for everything and everything in its place." Compare with "It's a great day for singing the blues."

  More generally, the apostrophe is never used to denote the possessive of a pronoun: the correct forms are "its," "hers," "his," and "theirs."

- **Latin Abbreviations**  By these we mean

  cf., e.g., i.e., n.b., q.v.

  and the like. These are abbreviations for specific Latin expressions: *confer* (compare), *exempli gratia* (for example), *id est* (that is), *nota bene* (note well), *quod vide* (which see). They have particular meanings, and you should strive to use them accurately. In particular, "cf." is often misused to mean "see." It actually means "compare." Sometimes "e.g." and "i.e." are interchanged in error; the first of these means "for example," and the second means (literally) "the favor of an example" or (more familiarly) "for the sake of example." It is difficult to use "n.b." with grace. If you are unsure, then use the English equivalent of which you *are* sure.

  In fact it is difficult to make a compelling case for "i.e." in favor of "that is," or for any of the other Latin substitutes in favor of their English equivalents. The punctuation and font selection questions connected with these Latin abbreviations are tricky (see [Hig] or [Fow] or [Chi] or [SaK]). For instance, many people do not know that you are always supposed to put a comma after "i.e." or "e.g.." To repeat, use these items with care.

- **Lay** and **Lie** "Lay" is a transitive verb and "lie" is intransitive. This means that "lay" is an action that you perform on some object, while "lie" is not. For instance, "Lay down your weary head," "Now I will lay down the law," and "I shall lay responsibility for this transgression at your feet"; compare with "I am tired and I shall lie down" or "Let sleeping dogs lie." Note, however, that the past tense of "lie" is "lay." Therefore you may say "Yesterday I was so tired that I laid down my books and then I lay down."

- **Less** and **Fewer** How many times have you been in the grocery store and gravitated toward the line labeled *Ten Items or Less*? Of course what is intended here is *Ten Items or Fewer*, and I have a special place in my heart for those few grocery stores that get it right. The word "fewer" is for comparing two numbers while "less" is for comparing quantity.[4] In mathematics we certainly say "3 is less than 5," and we do so because the discrete set $\mathbb{Z}^+$ is embedded in the continuum $\mathbb{R}^+$. You would never say "I have fewer milk than you do," just because milk is a continuum.

  You should also be wary of "smaller," which designates not number but size. Avoid saying "3 is smaller than 5," because "smaller" is a word about *size*: perhaps the digit 3 is smaller than the digit 5. It *could* also be correct to say "'5' is smaller than '3'" if comparison of digit size is what you intended: 5 versus 3.[5]

- **Lists Separated with Commas** (**the Serial Comma**) When you are presenting a list, separated with commas, then you should put a comma after every item in the list except the last. For example, say "the good, the bad, and the ugly" rather than "the good, the bad and the ugly." A moment's thought reveals that the former conveys the intended meaning of these distinct individuals; the latter may not, for the reader could infer that "bad" and "ugly" are simply two complementary descriptions of the same individual.

- **Numbers** Some sources will tell you that (whole) numbers less than 101 should be written out in words; larger numbers should be expressed in numerals. (Other sources will put the cutoff at twenty or some other arbitrary juncture.) A detailed discussion appears in [SG]. Such considerations are, for a mathematician, next to nonsensical. The main thing, and this advice applies to spelling and to many other *choices*, is to select a standard and to be consistent.

- **Obviously, Clearly, Trivially** These words have become part of standard mathematical jargon. This is too bad. In the best of circumstances,

---

[4]Another way to think about the matter is that "fewer" is used to compare discrete sets while "less" is used to compare continua.

[5]Note the extra single quotation marks to tell the reader that we are talking about the *digits* rather than the numbers.

when one uses these phrases, he/she is endeavoring to push the reader around. In the worst of circumstances, he/she is throwing up a smoke screen for something that he/she himself/herself has not thought through. It would be embarrassing to count the number of major published mathematical errors that have been prefaced with "Obviously" or "Clearly" or "Trivially." (No doubt the supreme deity's way of reminding us that "Pride goeth before the fall.") The indiscriminant use of these words is one of the ways that we have of kidding ourselves.[6]

As you proofread your manuscript relentlessly, and endeavor to weed out superfluous words, pay particular attention to the use, abuse, and overuse of these trite words. They add nothing to what you are saying, and are frequently a cover-up.

- *Overused Words* Many other words in the English language are also grossly overused. Among these are "very" and "most" and "nice" and "interesting." It is certainly very pleasant and most insightful to express great appreciation for a very nice and supremely interesting theorem; but I encourage you not to do so—at least not with these banal words. If such language represents how you wish to express yourself, then perhaps you have nothing to say. Instead think carefully about the real substance of what you are endeavoring to convey, and then find the substantive vocabulary to express it.

  Be aware that the language is littered with overused expressions that come into and out of fashion. The words "awesome," "totally," "dude," and "righteous" are current examples. The phrase "today I'm not 100%," foisted upon us by some semiliterate sports announcer, is currently the bane of our collective existence. Each field of mathematics has its own set of stock phrases and tiresome clichés. Endeavor not to propagate them.

  A good general principle is to put every word in every sentence under the microscope: What does it add to the sentence? Will the sentence lose its meaning if the word is omitted? Can the thought be expressed with fewer words? Strunk and White [SW] have a splendid discussion of the concept of weighing each word.

- *Plural Forms of Foreign Nouns* We all grind our teeth when we hear our freshmen say "And this point is the *maxima* of the function." To no avail we explain that "maxima" is plural, and "maximum" is singular. Yet we make a similar error when we do not differentiate "data" (plural) from "datum" (singular) and "criteria" (plural) from "criterion" (singular). To be sure, Latinisms take on a life of their own when transferred

---

[6]In a moment of exasperation, a friend of mine said of her soon-to-be-ex-husband, a mathematician, "You look at anything and you either say that it is 'very interesting' or 'trivial'."

to English. "Data," for example, can be construed as a collective singular noun. Similar ambiguities attend "agenda." This is a complicated business, because the word "data" is sometimes used as a collective noun and, in that context, is singular. Similar comments apply to "agenda" (plural to "agendum") and a few other words. In general, you should opt for the historically correct Latin forms. Above all, be consistent. As usual, exercise special care when dealing with foreign words.

- *The Possessive* When you express the possessive of a singular noun, always use 's. Thus you should say "Pythagoras's society," "the dog's day," "Stokes's theorem," "Bliss's book," "baby's bliss," and "van der Corput's lemma." The terminal "s" is omitted when you are denoting the possessive of a plural noun: "the boys' trunk," "the dogs' food," "the students' confusion."

  "Collective nouns" are treated in a special manner. For instance, we write "the people's choice" and "the children's folly": even though the nouns are plural, we denote the possessive *with* a terminal "s."

  Just because we frequently see such misuse in advertising and other informal writing, we sometimes get sucked into using extraneous apostrophes. Above all, remember that the apostrophe signals possession. Do *not* get bogged down with extraneous apostrophes. As an example, one often sees expressions like "This sentence contains a lot of TLA's." Here a TLA is a "three-letter acronym." What is wrong with this sentence? The last "word" in the sentence is supposed to be a plural—*not* a possessive. So the apostrophe is out of place. It should be "TLAs," not "TLA's." Likewise, do not write, "I surely miss the 1960's." It should be "1960s,", *not* "1960's."

- *Precision and Custom* At times, the goal of precision in writing flies in the face of custom. Antoni Zygmund once observed that the World Series of American baseball might more properly be called the "World Sequence." I am inclined to agree (in no small part out of fealty to my mathematical grandfather), but I must be over ruled by custom: if you use the phrase "World Sequence," then nobody will know what you are talking about. Bear this thought in mind when you are tempted to invent new terminology or new notation (see also the remarks in Section 2.4 on terminology and notation).

- *Principal vs. Principle* These words are easily confused. "Principal" means the main or the significant choice. "Principle" is a tenet or point that you want to make. Thus you speak of a "principal investigator" on a grant but the "localization principle" in partial differential equations.

- *Subject and Verb, Agreement of* Make sure that subject and verb agree in your sentences. A mismatch not only grates on the sensitive ear, but can seriously distort meaning. Consider the example "The set of all morphisms are compact." This syntax is incorrect. The *subject* (i.e., the person or thing performing the action) in this sentence is *set* (which is a singular noun). We should conjugate the verb "to be" so that it agrees with this subject. The point is that a singular subject requires a singular verb and a plural subject requires a plural verb. As a result, the grammatically correct statement is "The set of all morphisms *is* compact." (Note, in passing, that the original form of the sentence might have misled the reader into thinking that the writer was—rather clumsily—discussing a collection of compact morphisms.)

Of course the test is easy: omit the prepositional phrase "of all morphisms" and analyze the root sentence. Clearly "The set is compact" is correct while "The set are compact" is not. You will find the device of focusing on the root statement, or breaking into pieces (see our analysis of Subject and Object below), to be a valuable tool in analyzing many grammatical questions. Another way to look at the matter is that prepositional phrases can cause confusion.

As a parting exercise, consider the phrases "the sequence $\{z_n\}$ *converges* to $p$" while "the numbers $z_n$ *converge* to $p$." Think carefully about why both statements are correct.

- **Demonstrative Pronouns: *This* and *That*** We often hear, especially in conversation, phrases like "Because of this, we decided that." If we exercise the full force of logic, then we must ask " 'Because of' *what*?" and " 'we decided' *what*?" And this niggling query raises an entire body of common errors that I would like to point out. This corpus is not composed so much of errors in English usage, but rather errors in logic and precision. Consider the following examples:

    Shakespeare was an important writer. This tells us a lot about English literature.                    ✠

    A triangle is a three-sided polygon. This means that . . .                    ✠

    The day was bright and beautiful. Because of this, Mary smiled.
    ✠

    In each of these examples, my objection is "'this' what?" (Notice that I did *not* say "In each of these, my objection is . . . ." I was careful to say *precisely* what I meant.) The following passages convey the same spirit as the preceding three, but they actually *say* something:

    Shakespeare was an important writer. The forms of his plays and poems as well as his use of language have had a strong influence on English literature.

> A triangle is a three-sided polygon. The trio of sides satisfies the important *triangle inequality*.

> The day was bright and beautiful. Observing the weather caused Mary to smile.

> Here is a delightful example that was contributed by G. B. Folland:

> Saddam Hussein was determined to resist attempts to force Iraqi troops out of Kuwait, although George Bush made it clear that he did not want to be seen as a wimp. This caused the Gulf War.

If you were to ask someone to which clause "This" refers, then the answer you received would probably depend on that person's politics.

The message here is fundamental: as a default, do not use "this" or "that" or "these" or "those" without a clear point of reference. When the occurrence of "this" or "that" is fairly close to the referent, then the intended meaning is often clear from context. When instead the distance is greater (as in Folland's example), then confusion can result.

Repetition is a good thing, so repeat your nouns rather than refer to them with a potentially vague pronoun. There *will* be cases where the casual use of "this" or "that" is both natural and appropriate, but such instances will be exceptions. As a general rule, repeat your nouns.

Copy editor Rosalie Stemer says that a hallmark of good writing is that it answers more questions than it raises. Applying this philosophy will lead naturally to many of the points raised in this book, including the present one.

- **Where** One of the most common types of run-on sentence in mathematics is a statement with a dangling concluding phrase such as "where $A$ is defined to be ...." An example is

> Every convex polynomial function is of even degree, where we define a function to be convex if ...          ✠

We see this abuse so often that we are rather accustomed to it. This is also an easy crutch for the writer: he/she did not bother to plant the definition before this statement, so he/she just tacked the definition onto the end.

This practice is sloppy writing and there is no excuse for it: before you use a term, define it. You need not use a formal, displayed definition. But you must put matters in logical order. The example I have given is quite trivial; but in serious mathematical writing it is taxing on the reader to have to pick up definitions on the fly. Especially if you are writing with a computer, it is very easy for you to scroll up and put the needed definition where it belongs.

- ***Who* and *Whom;* Subject and Object**

Here we discuss relative pronouns.

Be conscious of the difference between "who" and "whom." The word "whom" is an object; used properly, it denotes a person who is *acted upon.* Put in other words, "who" is subjective, but "whom" is objective. The word "who" acts as the subject of a clause; the word "whom" acts as a direct object or as the object of a preposition. An example of the common misuse of the word "whom" is

The pastor, whom expected a large donation, smiled warmly.

Here the issue is what is the correct subject to put in front of the verb "expected." The word "whom" cannot act as a subject. The correct word is "who": "The pastor, who expected a large donation, smiled warmly." In the same vein, it is correct to say "To whom am I speaking?" and "Is he the man who was awarded the Nobel Prize?"

Also do not confuse "I"—used for the subject of a verb—and "me." The latter is an object, the former not. For example, "The teacher was addressing Bobby and I" is plainly wrong, since (here "I" is used incorrectly as the object of) the verb "addressing" calls for a direct object—"me." President Clinton's famous misstatement "Give Al Gore and I a chance to bring America back" is a dreadful error; nobody would say "Give I a chance . . . ." That sort of sentence analysis—breaking a sentence down to its component parts—is the method you should use to detect the error. The sentence

Him and me proved the isotopy isomorphism theorem in 1967.
✠

is an abomination. Unfortunately, even smart people make mistakes like this. Anyone can see that "Him proved the isotopy isomorphism . . . " and "Me proved the isotopy isomorphism . . . " are incorrect. But, somehow, the ganglia are more prone to misfire when we put the two sentences together. Conclusion: test the correctness of a sentence with compound subject (or any compound element) by breaking it into its component sentences.

# 1.11 More Rules of Grammar, Syntax, and Usage

## 1.11.1 Introduction

Here I include additional rules of grammar and syntax that are dear to my heart. They come up frequently in general writing, less so in specifically mathematical writing. They should prove useful in your expository work, and sometimes in your research work as well.

## 1.11.2 More Rules of Grammar and Syntax

- *Adjectives vs. Adverbs* An adjective is designed to describe, or to modify, a noun. An adverb is designed to describe, or to modify, a verb. Correct is to say

  This is a good book.

  and

  This is an expensive car.

  and

  The quick, brown fox jumped over the stupid, lazy dog.

  because "good," "expensive," "quick," "brown," "stupid," and "lazy" are adjectives. They modify the nouns "book," "car," "fox" (twice), and "dog" (twice), respectively. You may also say

  She shouts loudly.

  and

  He sings beautifully.

  and

  She strove sporadically to master her homework thoroughly.

  because "loudly," "beautifully," "sporadically," and "thoroughly" are adverbs. They modify the verbs "shouts," "sings," "strove," and "to master." Learn to distinguish between adjectives and adverbs, and learn to use both correctly.

  A nice example of the principles discussed here is

  I want to speak good English because I want people to think that I speak English well.

Here "good" is an adjective modifying the noun "English" while "well" is an adverb modifying the verb "speak."

After Paul Halmos had seen an early draft of the first edition of this book, he sent me the message "You write good." One can guess effortlessly that he was joking mischievously about this silly, little book. It may be noted, however, that "good" *can* be a noun. Consider, for instance, the sentence

The good that men do is oft interred with their bones.

We close this piece of advice by noting the very common problem with "good" and "well." "Good" is *always* an adjective. "Well" can be either a verb or an adjective. It makes sense to say "I feel well," because the adverb "well" modifies the verb "feel." It does *not* make sense to say "I feel good." Just because good is an adjective. But you *can* say "I am good."

- *Alternate* **vs.** *Alternative*   The adjectives "alternate" and "alternative" have traditionally different meanings, though they are often, and erroneously, used interchangeably. The word "alternate" (most commonly used in the form "alternately") refers to some pair of events that occur repeatedly in successive turns; the word "alternative" refers to a choice between two mutually exclusive possibilities. For example:

  Pierre alternately dated Mimi and Fifi. He had considered monogamy, but had instead chosen the alternative lifestyle of a concupiscent lothario.

- **The Verb** *To Be*   The verb "to be" is a linking verb; it implies a state of being, and can never take an object. Probably you have been hearing this assertion all your life. What does it mean?

  When you formulate the sentence

  I hit the ball.

  then "I" is the subject (of the verb "hit") and "ball" is the object (of the verb "hit"). But when you formulate the sentence

  I am the walrus.

  then "I" is the subject (of the verb "to be," conjugated as "am"), but "walrus" is the *predicate nominative* (also sometimes called the *predicate noun* or *subjective complement*). The word "walrus" is *not* receiving any action; it is simply restating or describing the subject (which is "I").

  When you are using nouns, such as "ball" and "walrus," you are unlikely to run into serious difficulties. With pronouns, however, you may. For you must carefully distinguish between "I"—the nominative singular used for *subjects*—and "me"—the accusative singular used for objects and also objects of prepositions. Likewise, you need to differentiate "we" (the nominative plural) from "us" (the accusative plural).

Thus it is technically incorrect (though rather common) to answer the query (over the telephone) "Is this Napoleon Bonaparte?" with the answer "This is me." This is because "is" requires a predicate nominative, *not* an accusative. Hence the correct rejoinder is "This is I" or "This is he."

To make a long story short, your writings should not include the statement "The person who proved Fermat's Last Theorem is me." Grammatically correct is "The person who proved Fermat's Last Theorem is I" or "It is I who proved Fermat's Last Theorem" or "I am the one who proved Fermat's Last Theorem." You should not, however, pen any of these statements unless you are Andrew Wiles.

- ***Compare*** and ***Contrast*** The words "compare" and "contrast" have different meanings. One compares two or more items in order to bring out their similarities; one contrasts two or more items in order to emphasize their differences. For instance, we can compare groups and semigroups because they are both associative. We can contrast them because one contains all inverse elements and the other need not.

- ***Different from*** and ***Different than*** The phrase "different from" is generally preferable to the phrase "different than." Notice that "from" is a preposition while "than" is a conjunction.

Examples are

His view of grammar is different from mine. $(*)$

and

His syntax is different from what I expected. $(**)$

Modern usage (see [Fra, p. 266]) suggests, however, that "different than" is permissible when it introduces a new clause. Thus, in the sentence $(**)$, you could instead say

His syntax is different than I expected.

You will have to decide which usage you prefer, but do be consistent.

- ***Due to*** Mathematicians commonly use the phrase "due to," and we often use it incorrectly. We sometimes say "due to the fact that" when instead "because" will serve nicely. The phrase "due to" tempts us to wordiness best resisted.

- **Farther** and **further** It is common to interchange the words "farther" and "further," but there is a loss of precision when you do so. The word "farther" denotes distance, while "further" suggests time or quantity. For example, one might say "I wish to study *further* the question of whether Lou Gehrig could throw the baseball *farther* than Ty Cobb."

- **Good taste and good sense** Suit your prose to the occasion. The writer of a Harlequin romance novel might write

  > Clutched in the gnarled digits of the syphilitic Zoroastrian homunculus was a dazzling Fabergé egg.      ✠

  while Raymond Chandler would have written something more like

  > The dwarf held a gewgaw.

  In mathematics, simpler is usually better. Flamboyant writing is out of place.

- **Hopefully** and **I hope** With due homage to Edwin R. Newman [New], I note that it is incorrect (at least in my view) to use "hopefully" (at the beginning of a sentence) when you mean to say "It is hoped that" or "I hope." The word "hopefully" is an adverb. It is intended to modify a verb. For example, consider the sentence

  > She wanted so badly to marry him, and she looked at him hopefully while she waited for a proposal.

  Note that the word "hopefully" modifies "looked." It is incorrect to say

  > Hopefully the weather will be better today.      ✠

  because the weather cannot hope. What you mean to say, of course, is

  > I hope that the weather is better today.

  By the same token, do not say "This situation looks hopeful." People can be hopeful, objects or things never.

  Monty Python tells us that "Mitzi was out in the garden, hopefully kissing frogs." If you are comfortable with the common misuse of "hopefully," then you will probably misunderstand this sentence.

  Actually, things are a bit more complicated than we have just indicated. In modern usage, an adverb can be used to modify almost anything except a noun. There is a concept of a *sentence adverb* that can modify an entire phrase. For example,

Frankly, my dear, I don't give a damn.

Regrettably, the meeting had to be cancelled.

Presumably he will now get the job.

In the first of these examples, the speaker (presumably Rhett Butler in the movie *Gone with the Wind*) is not giving frankly. He is in fact *not giving a damn* frankly. Similar remarks apply to the other two examples.

The reference [KnLR, p. 57] offers a detailed analysis of the history of the word "hopefully," and another, more liberal, point of view about its use. See also [BMW].

• *Infinitives, Splitting of* As a general rule, do not split infinitives. In other words, do not place an adverb between "to" and an inflected form of a verb.

For example, do not say "He was determined to immensely enjoy his food, so he smothered it in ketchup." The correct version (though one may argue with the sentiment) is "He was determined to enjoy his food immensely, so he smothered it in ketchup." Here the infinitive is "to enjoy" and the two words should not be split up. Curiously, the reason for this rule is an atavism: some of the languages that contributed to the formation of modern English, such as Latin and French, combine these two words into one. Our rule not to split the infinitive carries on that tradition.

In fact, it is somewhat misleading to say that it is a *rule* not to split infinitives. A perhaps more accurate statement is that many readers and writers find split infinitives to be grating on the ear.

There are a number of opinions on this matter. The "modern" point of view is that it is acceptable to split an infinitive when it sounds right; otherwise, it is not. For example, sometimes a mathematical sentence will resist the suggested rule. G. B. Folland supplies the example "Hence we are forced to severely restrict the allowable range of values of the variable $x$." Strictly speaking, the word "severely" splits the infinitive "to restrict." But where else could you put "severely" while maintaining the precise meaning of the sentence?

• *In terms of* Sentences of the form

Who is he, in terms of surname?        ✠

and

How is she doing, in terms of her math classes?                    ✠

are simply dreadful. Usually the phrase "in terms of" is gratuitous, and can be omitted entirely. Consider instead

What is his surname?

and

How is she doing in her math classes?

As English speakers, we often rely on quite meaningless idioms. As writers, we should try to avoid these pitfalls. Expressions such as "at this point in time" and "in terms of" are gratuitous and excessively wordy. Neither idiom adds any substance to your meaning. Instead of saying, "At this point in time we will consider . . . ," instead say "Now we will consider . . . ." Instead of saying "How is your food, in terms of tastiness?" instead say "How does your food taste?"

- **Need Only; Suffices to**    In written mathematics, we often find it convenient to say "We need only show that . . . " or "It suffices to show that . . . ." These are lovely turns of phrase. Strive not to overuse them, or to misuse them. Too often we see instead "We only need to show that . . . " or "Suffice it to show that . . . ." With these alterations, the message still comes across—but in a more halting and less compelling manner.

- **Parallel Structure**    The principle of parallel structure is that proximate clauses which have similar or related content and purpose are (often) more effective if they have similar form. The use of parallel structure is an advanced writing skill: good writing can be made better, more forceful, and more memorable with the use of parallel structure. Consider the dicta

Candy is dandy, but liquor is quicker.

or

Virtue is good, but sin is more fun.

Whether you approve of the sentiment or not, the first thought is memorably expressed—using a quintessential example of parallel structure. The second is somewhat parallel, but less so. As an exercise, try expressing the thoughts with more desultory prose, and see for yourself what is lost in the process.

The first inspirational quotations (from Sir Francis Bacon) in Chapters 3 and 5 provide less frivolous examples of parallel structure.

- **Dangling and misplaced modifiers**    Dangling and misplaced modifiers are a frequent cause for discomfort. For example,

> Shining like the sun, the man gazed happily upon the heap of gold coins.   ✠

The participial phrase "shining like the sun" modifies "man," whereas it should modify "the heap of gold coins." Better would be

> The man gazed happily upon the heap of gold coins, which shone like the sun.

Harold Boas contributes the following useful maxim: "When dangling, don't use participles."

Just as common, and even sillier, than misplaced modifiers are the dangling sort. To wit:

> By giving daily quizzes, the students' grades dramatically improved.

Who is giving daily quizzes here? The professor, of course. But he/she does not even appear in the sentence! Of course the phrase "By giving daily quizzes" requires a subject. The thought is expressed much more clearly as "By giving daily quizzes, the professor helped the students to improve their grades."

Be aware that infinitive phrases can also dangle. For example, "To get tenure, her mathematics must be superb." Here it is *not* the mathematics that gets tenured; it is the professor. A much clearer phraseology is "If she wants to get tenure, then her mathematics must be superb."

- ***Prepositions, Ending a Sentence with*** Many people object to ending a sentence with a preposition (words such as "to," "at," "of," etc.). Rather than say "Where do we stop playing at?" the purists suggest "At what point do we stop playing?" Better still is "When do we stop playing?" Rather than say "What book are you speaking of?" opt instead for "Of which book do you speak?" or "Which book is that?"

Like objections to the split infinitive, the distaste for preposition-ending sentences probably derives from a desire to apply Latin rules to English. Yet it is precisely because of this generalized distaste that you should at least be wary of this usage.

Often, when you are ending a sentence with a preposition, what is in fact occurring is that the errant preposition is a spare word—not needed at all. The preceding examples, and the suggested alternatives, illustrate the point.

Above all, try to keep your syntax simple and easily readable. Sometimes a sentence ending with a preposition sounds much more natural than a sentence that goes into contortions trying to avoid this problem.

An old joke has a yokel trying to find his way across the Harvard campus. A Brahmin student corrects him sternly for posing the question "Excuse me. Where's the library at?" After the Harvardian explains at length that one does not end a sentence with a preposition, the yokel tries again: "Excuse me. Where's the library at—*jerk*?" The yokel here is making a good point: Grammatical rules, if stilted, are best ignored.

As an exercise, find a better way to express the following sentence (which ends with five prepositions, and which I learned from Paul Halmos by way of [KnLR]):

What did you want to bring that book I didn't want to be read to out of up for?

Harold Boas cautions: "Watch out for prepositions that sentences end with."

- *Quotations*   We do not often include quotations in mathematics papers. If you decide to include a quotation, then be aware of the following technicality. Logically, it makes sense to write a sentence of the following sort:

As Methuselah used to say, "When the going gets tough, the tough get going".          ✠

What is logical here is that the quotation itself is a proper subset of the entire sentence; therefore it stands to reason that the terminal double quotation mark should occur *before* the period that terminates the sentence. Unfortunately, logic fails us here. Admittedly typesetters are still debating this point, but the current custom in the United States is to put the period *before* the closing double quotation mark. Open any novel and see for yourself. Thus the sentence *should* be written

As Methuselah used to say, "When the going gets tough, the tough get going."

Like periods, commas should also be placed inside the quotation marks: "When the going gets tough," as Methuselah used to say, "the tough get going."

By the rules of *American* usage, commas and periods should be placed inside quotation marks, and colons and semicolons outside quotation marks (see [SG, p. 222] and [Dup, p. 192]). Surprisingly, perhaps (but logically), colons and semicolons should be placed *outside* the quotation marks. Placing exclamation points and question marks inside or outside of quotation marks depends on context. British usage is even more ambiguous. This is all a bit like the infield fly rule in baseball. But do be consistent, and be prepared to arm-wrestle with your publisher or with your copy editor if you have strong opinions in the matter.

If your quotation is $n$ paragraphs in length, then there is an opening double quotation mark on every paragraph. There is no closing double quotation mark on paragraphs 1 through $(n-1)$; but there certainly *is* a closing double quotation mark on paragraph $n$. Again, check any published novel to see that this is the case.

- **Redundancy** Logical redundancy, used with discretion, can be a powerful teaching device. By contrast, avoid (local) verbal redundancy. The phrases "old adage," "funeral obsequies," "refer back," "advance planning," "strangled to death," "invited guest," "body of the late," and "past history" display an ignorant and superfluous use of adjectives. Avoid constructions of this sort.

- **Shall** and **Will** In common speech, the words "shall" and "will" are often used interchangeably, or according to what appeals to the speaker. In formal writing, traditionalists note a distinction: when expressing belief regarding a future action or state, "shall" is used for the first person ("I" or "we") and "will" is used for the second person ("you") or third person ("he," "she," "it," or "they"). To express determination, the first person could use "will." These rules, taken from [SW], are illustrated whimsically in that source by

  **Bather in Distress:** "I shall drown and no one will save me."

but

  **Suicide:** "I will drown and no one shall save me."

In practice, these distinctions are largely lost in modern American usage. "Will" has become the general all-purpose choice for most people. Still, there are instances in which both "shall" and "will" can be used very effectively—if not for differences of meaning then certainly for differences of sound and emphasis.

A more modern example of careful usage of "shall" and "will" comes from President Lyndon Johnson:

  I shall not seek, nor will I accept, the nomination of my party.

- **That** and **Which** The relative pronoun "that" is used to denote *restriction*, while the relative pronoun "which" denotes *amplification*. For example, "I am speaking of the vase that sits on the table" and "The book that is by Gibbons is in the study." Compare with "The vase, which is red, sits on the table" and "The book, which is by Gibbons, is fascinating."

In these instances, "that" introduces essential information: the vase that is on the table, *not* the one that is on the floor; the book by Gibbons, *not* the one by Pascal. By contrast, "which" introduces descriptive but nonessential information: the vase is on the table and the book is fascinating; and, by the way, the vase is red and the book is by Gibbons. It is worth noting that nonrestrictive clauses are set off by commas, while restrictive clauses are not (see the examples in the last paragraph).

In mathematics, the difference between "that" and "which" can sometimes be crucial. Consider this example:

> A holomorphic function on a connected open set that vanishes on $S$ must be identically zero.

Compare with

> A holomorphic function on a connected open set which vanishes on $S$ must be identically zero.

Which is correct? Think about the logic. What we are saying is that a holomorphic function $f$ on a connected open set such that $f(z) = 0$ for $z \in S$ must be identically zero. (For the mathematics, note that, in one complex variable, a set $S$ with an interior accumulation point will suffice for the truth of the statement.) Phrased in this way, the statement is restrictive: a holomorphic function with a certain additional property must be zero. Thus the correct choice is "that" rather than "which."

Modern grammarians approve of the use of "which" for "that" in suitable contexts. Consult a grammar book, such as [SG], for the details.

I have already noted that it is sometimes useful to let your ear overrule the strict code of grammar. In particular, there are times when "which" sounds more weighty, or more formal, than "that." Thus some writers will make the technically incorrect choice, just to achieve a certain effect.

As already noted, the rules of grammar and syntax are not absolute. English usage is constantly evolving. While some current aspects of usage are fads and nothing more, others become common and are finally adopted by the best writers and speakers. Those tend to stay with us. But there is a more subtle point. Sometimes a sentence formed according to the strict rules of usage *sounds awkward*. A classic example (usually attributed to Winston Churchill) is

That is the sort of behavior up with which I will not put.                    ✠

Notice that the speaker is going into verbal contortions to avoid ending the sentence with a preposition. The result is a sentence so absurd that it defeats the main purpose of a sentence—to *communicate*. Better is to say

That is the sort of behavior that I will not put up with.

While technically incorrect—because the preposition is at the end of the sentence—this statement nevertheless will not grate on the ears of the listener, will convey the sentiment clearly, and will get the job done. Of course it would be even better to say

I will not tolerate that sort of behavior.

This sentence conveys exactly the same meaning as the first two. But it has the advantage that it is direct and forceful. In most contexts, the last sentence would be preferable to the first two. This is again a matter of thinking about what the message is intended to be. And here is a point that I will make several times in this book: often it is a good idea *not* to wrestle with a sentence that is not working; instead, reformulate it. Make it sound more natural. We did precisely this with the last example.

# Chapter 2

# Topics Specific to the Writing of Mathematics

*What I really want, doctor, is this. On the day when the manuscript reaches the publisher, I want him to stand up—after he's read it through, of course—and say to his staff: "Gentlemen, hats off!"*

Albert Camus
*The Plague*

*You don't write because you want to say something; you write because you've got something to say.*

F. Scott Fitzgerald

*So I'm, like, "We need to get some food." And he's, like, "I don' wanna go th' store. How 'bout some 'za?" And I'm, like, "Well, we gotta eat, dude. I could get like totally into a pizza." And he's, like, "No biggie." And I'm, like, "This guy is grody to the max. Gag me with a spoon."*

A Valley Girl

*We have read your manuscript with boundless delight.*
*If we were to publish your paper,*
*it would be impossible for us to publish any work of lower standard.*
*And as it is unthinkable that in the next thousand years*
*we shall see its equal, we are, to our regret,*
*compelled to return your divine composition and to beg*
*you a thousand times to overlook our short sight and timidity.*

Memo from a Chinese Economics Journal
From *Rotten Rejections* (1990)

*Having imagination, it takes you an hour to write a paragraph that, if you were unimaginative, would take you only a minute. Or you might not write the paragraph at all.*

Franklin P. Adams
*Half a Loaf* (1927)

The advice in this chapter is intended primarily to apply to the writing of a mathematical research paper. But some of the ideas are more universal than that. The thoughts would also certainly apply to the writing of a monograph, and in some contexts to expository writing as well.

Modern technology enables a marvelous writing environment—at least in principle. If, for example, I am a `Windows`®️ user, then I can have my text editor going in one window (this is where I actually do my writing), a thesaurus and dictionary on optical disk in another, the library's online catalog in another, and `MathSciNet` online in a fourth. Passing from one environment to the next requires only a keystroke or a mouse click or two. Clearly such an environment makes tedious trips to the library a thing of the past, and makes assembling a bibliography relatively quick and easy.

## 2.1   How to Organize a Paper

### 2.1.1   Title, Authors, Abstract, Etc.

To begin, a mathematics paper has certain technical components. It requires a title, and that title should convey some information to the reader. If it does not, then the reader is likely to move on to ostensibly more stimulating reading matter, without looking any further at your work. A title like *On a theorem of Hoofnagel* says almost nothing. One like *On differentiation of the integral* is only slightly better; but at least now the reader knows that the paper is about analysis, and he/she has a rough idea what sort of analysis. The title *Quadratic convergence of Lax–Wendroff schemes with optimal estimates on the error term* is ideal. This title tells the reader exactly what the paper is about and, further, what point it makes.

Of course an equally important component of your paper is the identification of the author or authors. At the beginning of your career, pick a name for yourself and stick to it. And I do not mean a name like "Stud" or "Juicymouth." I might have called myself Steven George Krantz or S. G. Krantz or S. Georgie Krantz or any number of other variants. I chose Steven G. Krantz, just as it appears on the title page of this book. When an abstracting, indexing, or reviewing service endeavors to include your works, you want it to be a zero-one game: it should retrieve all your works or none of them. You do not want any to be left out, and you should leave no doubt as to your identity.

Here is a quick run-down of other technical components that belong in most papers:

(1) affiliations of authors,

(2) postal addresses and email addresses of authors,

(3) date of submission,

(4) abstract,

**(5)** key words,

**(6)** AMS Subject Classification numbers,

**(7)** thanks to granting agencies and others.

Items **(1)**, **(2)**, and **(3)** require no discussion; topic **(4)** is discussed in Section 2.5. Let us say a few words about **(5)**–**(7)**.

The key words are provided so that *Math. Reviews* and other archiving services can place your paper correctly into a database. You simply list the key words in a footnote: **Key Words:** *holomorphic, pseudoconvex, analytic continuation.* Endeavor to choose words that reveal what your paper is about; you want words that will definitely lead a potential reader to your paper. Thus "new," "interesting," and "optimal" are not good choices for key words. Instead, "pseudoconvex," "Cauchy problem," and "exotic cohomologies" *are* good choices.

Similar comments apply to the AMS Subject Classification Numbers. The American Mathematical Society has divided all of mathematics into about 98 primary classification areas (rather like *phyla* in the classification of animals)— numbered 00 to 97 (the AMS skips some of these numbers)—and these in turn into subareas. Assigning the correct classification numbers to your paper is a reliable way to put your paper before the proper audience. It also ensures that your paper is classified correctly. The `MathSciNet` website lays out the AMS Subject Classification scheme. Key words and classification numbers usually appear in footnotes on the first page of your paper; some journals specify instead that title, abstract, key words, and classification numbers appear on a separate "Title Page." You should check the Instructions to Authors for the particular journal that you are dealing with to learn how they want things done. If you are submitting the paper at a website, then these things are mostly taken care of automatically.

Finally, item **(7)**: often it will be appropriate to thank other mathematicians for helpful conversations or specific hints. Sometimes you will thank someone for reading an earlier draft of the paper, or for catching errors. Strictly speaking, you should ask a person before you thank him/her in public (because, for example, most people would not want to be thanked heartily in a paper that turned out to be hopelessly incorrect). However, as a matter of fact, most people do not engage in this formality; and most of those who are thanked do not object. Occasionally you may wish to thank the referee for helpful comments or suggestions (best is to do this *after* the paper has been refereed—not in advance, or in anticipation of a friendly referee); sometimes you will need the editor's help in handling this particular "thank you" correctly. Writers also often thank their spouses for forbearance, or their typist (if, as is increasingly rare, one has a typist) for a splendid job with the manuscript. Sometimes one thanks one's department for time off to complete the work, or for the opportunity to teach an advanced seminar in which the work may have been developed. The one particular form of thanks that you are honor bound to include is thanks to any agency—government, university, or private—that has provided you financial (or other) support. In some cases, this thanks is mandatory; in all cases, it is an eminently appropriate courtesy. These thanks usually go in a brief paragraph at the end of the Introductory section of the paper.

### 2.1.2 Organization of the Paper

Now let us turn to the contents of the paper. A mathematical paper is not an exercise in self-indulgence (in content it might be, but in form it definitely should not be). You are writing about a topic on which you have become expert. You have made an advance, and you want to share it with the mathematical community. You should maintain the perspective of a serious and competent communicator.

The simplest way to write a paper is to introduce some notation, state your theorem, and begin the proof (for simplicity I am supposing that this is a "one-theorem paper"). Such a procedure probably involves the least effort on your part, it gets the theorem recorded for posterity, and it might even get the paper published. But this methodology is less effective if you genuinely want your work to be read and understood, and if you want the ideas disseminated to the broadest possible audience.

Indeed, a good mathematics paper is *not* necessarily written in strict logical order. The reason lies in theories of learning due to Piaget and others. The point is simply this: While it can be useful—when recording mathematics for archiving in the literature—to develop ideas à la Bourbaki/Hilbert in strict logical order, *this is not the way that we learn.* It is not the way that a typical human being—even a mathematician—apprehends ideas. This is the case even if the reader is an expert in the subject, just like yourself. In fact we learn, generally speaking, by proceeding from particular examples to general ideas.

Reading a mathematics paper is hard work, and a typical reader approaches the task with caution. Most people will not read more than a couple of math papers per month—I mean *really* read them, verifying all the details. However, those same people will *look* at several dozen papers each month. We all receive a great many preprints over email or the Internet or by way of preprint servers. We must make choices about which ones to *read*.

Having established this premise, let us think about what sort of paper will encourage the potential reader to plunge in, and what sort will not. If the first couple of pages of the paper consist of technical definitions and technical statements of theorems, then I would wager that most potential readers will be discouraged. Imagine instead a paper written as follows. The first paragraph or two summarizes the main results of the paper, in nontechnical language. The next several paragraphs provide the history of the problem, describe earlier results, and state exactly what progress the current paper represents. This introduction concludes, perhaps, with acknowledgements and an outline of the organization of the paper (either in Table of Contents form or paragraph form).

A reader faced with the latter organizational form has many advantages. This person knows **(i)** what the paper is about, **(ii)** why the result of the paper is new, (iii) what is the context into which the paper fits, and (iv) whether he/she wants to read on.

One person whom you must keep in sharp focus as you craft your paper is the referee. You cannot, indeed you must not, assume that the referee will compensate for your shortcomings. If *you* do not explain what the paper is about, why you wrote it, why your theorems are new, why this paper makes an interesting contribution, why its techniques are original—then nobody else is going to do it for you. And the referee, who almost certainly will not read the *entire paper*, will (if the introductory portion of your paper is not up to snuff) conclude quickly that your masterpiece is just another piece of second-rate drivel and will reject it.

Back to the chase: Imagine that, having concluded the above-described introductory section of the paper (Section 1), you (the writer) turn to the necessary technical definitions and a formal statement of results. This central material would be the substance of Section 2 of the paper (assuming that the introductory material, discussed in the last paragraphs, was Section 1). Now the reader—the expert who has stayed with you this far—knows precisely what he/she is getting himself/herself into. Turning to Section 3, you (the writer) can now dive into all the nasty details of the proof. Right? Wrong.

Reading a difficult mathematical proof in strict logical order is an onerous task. If the first five pages of Section 3 consist of a great many technical lemmas and their proofs, with nary an indication of where things are going, of what is important, and of what is not, then many readers will be discouraged. Let me now describe a better way.

It is more work for the writer, but definitely a great favor to the reader, to organize the paper as follows. Section 3 should consist of the "big steps" of the proof. Here you should formulate the technical lemmas (provided that the reader can understand them at this point), and then you should describe how they fit together to yield the theorem. You should push the nasty details of the proofs of the lemmas to the end of the paper.

The proposed organizational scheme makes sense because it ensures readability. First, the reader can decide at each of these signposts how far he/she wants to get into the paper. Each new epsilon of effort on his/her part will yield additional and predictable benefit. And the hardest and most technical parts are left to the end for the real die-hard types. This writing style is of course beneficial for the reader; it will also aid you as the writer. It disciplines you and forces you to evaluate and pre-digest what you have have to say, and will tend to reduce errors.

The principles of writing a math paper that have been described here do not apply to every paper that is, has been, or ever will be written. They probably do not literally apply to the Feit/Thompson paper on the classification of finite simple groups (an entire issue of the *Pacific Journal*) or to Andrew Wiles's proof of Fermat's Last Theorem (an entire issue of the *Annals*). They certainly apply to a twenty-page, "one-theorem" paper. And the general principles described above probably apply in some form to virtually any mathematics paper.

And now a word about redundancy. In general, redundancy is a good thing.

Unfortunately, we mathematicians tend to think that when we have said some-thing once clearly, then that is the end of it; nothing further need be said. This observation explains why mathematicians so often lose arguments—because we are loath to repeat ourselves. You must repeat. On the one hand, we do not want to bore the reader by rehashing trivial details; on the other hand, we want to aid the reader by reminding him/her of the key ideas. Help the reader by recalling definitions—especially if the definition was given 50 pages ago. If you need to use the definition *now*, and if you have not used it for quite a while, then give the reader some help. Give a quick recap or at least a cross-reference; do likewise for a theorem or a lemma that you need to recall. Think of how much you would appreciate this assist if you were the reader.

## 2.2   How to State a Theorem

### 2.2.1   Efficient and Elegant Statement of Your Results

There are some mathematical subjects—geometric measure theory is one of them—in which the custom is for the statement of a theorem to occupy one or more pages, and for the enumerated hypotheses to number twenty-five or more. This practice is unfortunate because it makes the subject seem impenetrable to all but the most devoted experts, who are wont to claim that their subject prevents any other formulation of the theorems: this is just the nature of the beast. I would like to take this opportunity humbly to disagree. To be sure, you may need to expend a little extra effort; but you need never state a theorem in this opaque manner.

 You should strive to hold the statement of a theorem to fewer than ten lines, and preferably to no more than five lines. (Some books on writing assert that a theorem should consist of only a single sentence!) How can you do this if there are twenty-five hypotheses? First of all, the assertion that there are twenty-five hypotheses is only a manifestation of what is going on in the writer's mind. Mathematical facts are immutable and stand free from any particular human mind, but the way that we describe them, verify them, and understand them is quite personal. In particular, the way that a theorem is presented is subject to considerable flexibility and massaging. Let us consider a quick and rather artificial example:

**Theorem:** Let $f$ be a function satisfying the following hypotheses:

1. The function $f$ has domain the real number line;

2. The function $f$ is positive;

3. The function $f$ is uniformly continuous;

4. The function $f$ is monotone;

5. The function $f$ is convex;

6. The function $f$ is differentiable except possibly on a set of the first category;

7. The function $f$ has range dense in the positive real numbers;

8. The function $f$ has no repeated values;

9. The function $f$ is a weak solution of the differential equation

$$Lf = 0$$

(where the operator $L$ has been defined earlier in the paper);

10. The function $f^2$ is a subsolution of $Lf = 0$.

Then $f$ operates, in the sense of the functional calculus, on all bounded linear operators on a separable, real Hilbert space $H$.

This sample "theorem" has only ten hypotheses, and these assumptions are not all that difficult to absorb; but the format serves to illustrate our point. As we have formulated the theorem above, the reader will get bogged down in the hypotheses; he/she is liable to view the body of the theorem itself as almost an afterthought.

Here is a more efficient, and more user-friendly, manner in which to state the theorem.

## 2.2.2 A Better Way to State Your Result

Prior to the statement of the theorem, we formally define a function to be *regular* if it is defined on the real line, uniformly continuous, convex, monotone, and positive. Further, we define a function to be *amenable* if it has range dense in the positive reals and has no repeated values. Finally, let us say that a function $f$ is *smooth* if: **(i)** it is differentiable except possibly on a set of the first category, **(ii)** it is a weak solution of $L$ and, in addition, **(iii)** $f^2$ is a subsolution of $L$. Each of these pieces of terminology should be stated as a formal definition, prior to the formulation of the theorem. Moreover, we should state that, until further notice, $H$ will designate a separable, real Hilbert space and $\mathcal{L}(H)$ the bounded linear operators on $H$. With this groundwork in place, we can now state the theorem as follows.

**Theorem:** If $f$ is a regular, amenable, smooth function, then it operates on $\mathcal{L}(H)$ in the sense of the functional calculus.

Notice that, by planning ahead and introducing the terms "regular," "amenable," and "smooth," we have grouped together cognate ideas. We have stated the theorem in one sentence, and in just two lines.

We are not just engaging in sleight of hand; in fact, we are providing organization and context. We are also helping the reader by keeping the statement of the theorem short and sweet. The reader will come away from reading the theorem remembering that **(i)** there is a hypothesis about $f$ involving continuity, convexity, and so forth, **(ii)** there is a hypothesis about the value distribution of $f$, and **(iii)** there is a hypothesis about the way that $L$ acts on $f$. The conclusion is that $f$ operates on $\mathcal{L}(H)$. You, the writer, have done some of the work for the reader, and given him/her something to take away. The reader can always refer to the text for details as they are needed. But this new format is more readable than the first, where the reader is likely not to quite know what he/she has read, nor when and where he/she can use it.

Also note that we managed to state the theorem in one sentence, and in just two lines.

## 2.3   How to Prove a Theorem

### 2.3.1   The Proof Is in the Pudding

What I mean here, of course, is "how to *write the proof of a theorem*." You are not doing your job—unless the proof is short and fairly simple—to begin at the beginning and charge through to the end. A proof of more than a few pages should be broken into lemmas and corollaries (typographers call these *enunciations*) and organized in such a fashion that the reader can always tell where he/she has been and where he/she is going.

A useful device in writing up a proof is the "Claim." This tool is often used in the following manner. You have set up the basic pieces of your proof; you have defined the sets and functions and other objects that you need. You are poised to strike. Then you write "We claim that the following is true." Then having stated the claim, you say "Assuming this claim for the moment, we complete the proof."

Used correctly, this technique is a terrific psychological device. It allows you to say to the reader "This is the crux of the proof, but its verification involves some nasty details. Trust me on this for the moment, and let me show you how the crux leads to a happy ending." The reader, having arrived at the end of the proof (modulo the claim), will feel that progress has been made and he/she will be in a suitable mood either to study the details of the claim or to skip them and come back to them later.

Another useful trick—nearly logically equivalent to the "claim"—is to enunciate a technical lemma right at the point where you need to use it (to enunciate it well in advance would make almost no sense to the reader), but then to say "Proof deferred to Section 8." If you indulge in this trick, be sure that your paper is well organized and that the different parts of the paper are well labeled. Do not leave your poor reader with a head full of dangling claims and unproved lemmas to sort out. A good rule of thumb ([Gil, p. 8]) is to be sure that your reader always knows the *status* (is it true or false? has it been proved or not?) of every statement that you make.

## 2.3.2 Technical Stuff in the Back

In Section 2.1 I have advocated that a paper should be organized so that the technical material is at the back and the explanatory points at the front. An excellent example of this type of writing—indeed the place where I first learned of the technique—is [Ker]. The paper should proceed, by gradations, from the latter to the former. The proof of a theorem should proceed in roughly the same way. You, as the author and creator of the theorem, have the whole thing jammed into your head; it has no beginning and no end—it just resides there. Part of the writing process is to transfer this platonic mass from your head to someone else's. Thus, as you write, try to provide signposts so that the reader always knows how the explanations just offered lead logically to the next points, and then on to the climactic point when the theorem is finally proved. This writing goal is best achieved by pushing the technicalities to the end.

## 2.3.3 Announce Your Theorems and Proofs

Next we treat an important example of what *not* to do—and it is done far too often by well-meaning authors. Many books, and some papers, are written as follows: the author rattles on for several pages—chatting about various mathematical topics—and then abruptly says

> Note that we have proved the following theorem:
>
> **Theorem [The Riemann Mapping Theorem]** Let $\Omega$ be a simply connected, proper subset of the complex plane . . . . ✠

Good heavens! What a disservice to the reader. The Riemann mapping theorem is a milestone in mathematical thought, perhaps even in human thought. It is important for each theorem, but especially for a big theorem like the Riemann mapping theorem (RMT), to first state the result[1] and then to announce that you are beginning the proof by writing (or typesetting)

## Proof:

Each of the steps in the proof of the RMT—the extremal problem, the normal families argument, etc.—is a subject in itself. The writer must lay these milestones out for the reader and must pay due homage to each. The offhand "Note by the way that we have proved the Riemann mapping theorem" is a real travesty and ignores the author's duty to *explain*. Rise above the idea that it suffices for the writer to somehow record the key idea on the page; if you, the author, have not crafted them and worked them and, indeed, handed them to the reader, then you have not done your job.

And here is a small note about proofs by contradiction. Some mathematicians begin a proof by contradiction with

---

[1] Of course you can precede the enunciation of the theorem with some motivation and explanatory material.

> Not. Then there is a continuous function $f$ ...          ✠

Others begin with

> Deny. Then there is a continuous function $f$ ...          ✠

This is all rather cute; the first of these is perhaps a tribute to John Belushi and the *Saturday Night Live* gang. Reliance on monosyllables may seem acceptable by dint of the fact that it is common in modern culture. But both examples (and these are *not* made up—mathematicians actually have been known to write this way) hinder the task of *communicating*. A preferred method for beginning a proof by contradiction is

> Seeking a contradiction, suppose that $f$ is a continuous, real-valued function on a compact set $K$ that does not assume a maximum. Then ...

## 2.4   How to State a Definition

### 2.4.1   Definitions as Fundamental Units

Definitions are part of the bedrock of mathematical writing and thinking. Indeed, mathematics is almost unique among the sciences—not to mention other disciplines—in insisting on strictly rigorous definitions of terminology and concepts. Thus we must state our definitions as succinctly and comprehensibly as possible. Definitions should not hang the reader up, but should instead provide a helping hand, as well as encouragement for pushing on. Definitions should be presented in logical order and *before* they are actually needed or used.

As much as possible, state definitions briefly and cogently. Use short, simple sentences rather than long ones. To avoid excessively complex and self-referential definitions, endeavor to *build* ideas in steps. For instance, suppose that you are writing an advanced calculus book. At some point you define what a function is. Later you say what a continuous function is. Still later you state what the intermediate-value property for continuous functions is. Further on, you use the latter property to establish the existence of $\sqrt{2}$. You do not want all at once to spit out all these ideas in a single sentence or a single paragraph. In fact, you build stepping stones leading to the key idea, so that the reader can internalize idea $n$ before going on to idea $(n + 1)$.

Just how many definitions should you supply? If you are writing a paper on von Neumann algebras (algebras of bounded operators on Hilbert space), then you certainly need not say what a Hilbert space is, nor what a bounded linear operator is. Since every graduate student who has passed through the qualifying exams is familiar with these ideas, you may take the reader's awareness of the notions for granted (which is why we have qualifying exams). Define $\mathcal{L}(H)$ (see Section 2.2) only if you think that readers likely will misinterpret this (rather standard) notation. Of course you would have to define "regular," "amenable," and "smooth" (the terminology that we introduced in Section 2.2). Those terms are not standard, and have been given other specialized meanings elsewhere.

### 2.4.2  Supply Enough Definitions

There is a degree of subjectivity here. If your paper supplies too few, or poorly written, definitions, then both the referee and the readers will lose their patience. If your paper supplies too many definitions, then you also will irritate your audience. For standard terminology, you could give a well-known reference like Dunford and Schwartz [DS] or Griffiths and Harris [GH] or Birkhoff and MacLane [BM] or Kuratowski [Kur]. This habit is preferable to taking up valuable journal space with a rehash of well-known ideas.[2] Do not, however, rely upon a semi-obscure journal article for terminology. If this is your best reference for definitions, then you should probably repeat the definitions yourself.

To be sure, there is some terminology that you simply cannot take the space to repeat or define, even though it is rather advanced. For example, in a journal article you cannot rehash—for the convenience of your readers—the standard theory of elliptic partial differential equations, nor the basics of $K$ theory, nor the guts of the Atiyah–Singer Index Theorem. (In a book you in fact *can* indulge in such a review; I treat book writing elsewhere.) Try to refer the reader to a good source for the important ideas on which you are building.

### 2.4.3  Terminology Can Unify Ideas

Try to avoid introducing any more new terminology than is necessary. If your paper contains a plethora of unfamiliar language, then it may cause your reader to suspect that you actually have nothing to say. And if there is a standard bit of notation or terminology for what you are saying, then by all means use it. I once saw a paper in a standard mathematics journal of good repute that defined the space $Q_{\mathrm{reg}}^{17}$ to be the set of all bounded holomorphic functions on the unit disk in the complex plane. Of course, the well-known notation $H^\infty(D)$ describes this space of functions, and it is virtually mandatory to use *that* notation. The proposed alternative notation would be reasonable only if the author is introducing a whole new scale of function spaces in which $H^\infty$ arises in a natural way. If that were the case, then the author should certainly should have mentioned this relationship explicitly. (For example, all the standard function spaces—$L^p$, Lipschitz, Sobolev, Hardy, Besov, Nikol'skii, etc.—are special cases of the Triebel–Lizorkin spaces $F_p^{\alpha,q}$. Thus, in certain contexts, it would be appropriate to refer to the Lebesgue spaces, or the Sobolev spaces, using the Triebel–Lizorkin notation.)

When you do need to create new notation, bear in mind that it can be as important as a new theorem. As an example, the notation of differential forms is a small miracle. Large parts of geometric analysis would be completely obscure without it. Of course, you cannot perform at the level of Élie Cartan every time you dream up a piece of notation. But you can consider following these precepts:

---

[2]But do your reader the kindness of telling him/her to what *part* of [DS] or [GH] to refer. Best is to give a specific definition number. But giving the page number is also fine. Or even the section number is acceptable.

**(1)** do not create new notation if there already exists well-known notation suitable for the job at hand;

**(2)** if you must introduce new notation, then think about it carefully;

**(3)** strive for simplicity and clarity at all times.

In sum, experiment with several different notations before you make a final decision. Consult the standard references in the field to see whether they give you any ideas. If possible, try your new notation out on a colleague, or on one of your graduate students.

Finally, from a technical standpoint, a definition should almost always be formulated in "if and only if" form. For example

> A function $f$ on an open interval $I$ is said to be *continuous* at $c \in I$
> if and only if, for every $\epsilon > 0$, there is a $\delta > 0$ such that . . .

In practice, we generally replace the phrase "if and only if" in this definition with "if." We do so partly out of laziness, and partly because the "if" phraseology is less cumbersome than "if and only if." The price that we pay for this convention is that we must teach our students to read definitions; lazy or not, we *do not* write what we mean.

Although nobody will punish you for writing "if and only if" in your definitions, and some will appreciate it, it is usually best to follow mathematical custom and simply to write "if." A useful, and modern, compromise is to use Paul Halmos's invention "iff" (see Section 1.8). The word 'iff' captures the brevity of "if" but carries the precision of "if and only if."

## 2.5 How to Write an Abstract

### 2.5.1 What Is an Abstract?

Many journals now require that, when you submit a paper, you include an abstract of the paper. The abstract, usually not more than ten lines, should convey on a quick reading what is the paper's substance. According to the strictest standards, the abstract should be self-contained, should not make any bibliographic references, and should contain a minimum of notation and jargon.

A rough rule of thumb is that any reader who looks at your paper will read the abstract, only 20% of those will read the introduction, and perhaps one fourth of that 20% will actually dip into the body of the paper. This being the case, your abstract is obviously of pre-eminent importance. Many indexing and reviewing services will rely on your abstract. So it had better give a clear picture of what is in the paper.

As usual, endeavor to employ simple, short, declarative sentences in your abstract. Eschew nasty details. Do not say, with a plethora of $\epsilon$s and $\delta$s, exactly what interior elliptic estimate you are proving; instead state that you are proving a new interior elliptic estimate in the Nikol'skii space topology and that it improves upon classical results of Nirenberg. State that it has applications to certain free-boundary problems. The interested reader will then be motivated to move on to the introduction, where further details are provided.

### 2.5.2    What to Include in the Abstract

The journal and the archiving agencies will rely on your abstract to place your paper in the firmament. So you must not have bibliographic references in your abstract, and you must not have a lot of notation and fancy formulas. The abstract should be mostly words, and it should *describe* what is in the paper. It should not provide too much detail, but should give the reader a clear idea of what the paper is about. It should be fairly brief. Twenty lines is already too long.

If your abstract is too long or too short, then the journal editor will likely make you rewrite it. The "Instructions to Authors" section should give you an idea of what is required for an abstract in *that particular* journal. Study several abstracts in the journal to which you plan to submit to get an idea of what is suitable.

## 2.6    How to Write a Bibliography

### 2.6.1    Citing Your References

The bibliography, or list of references, is one of the most important components of a mathematical work. It is essential in research articles, in books, and in expository writing to give a thorough and complete accounting of your sources. The bibliography tells the reader where you are coming from and where you are going, it keeps you honest, and it provides critical assistance for those readers not already familiar with the subject.

Conscientious writers assemble their bibliographies only from primary sources. The book [Hig, pp. 87–8] has several examples of bibliographic inaccuracies in the literature that have been propagated for dozens of years because reference $(n + 1)$ was always copied from reference $n$. Put in other words, the writers failed to look at the original source; instead they simply copied other writers' references to it.

The book [Hig] also advises you never to retrieve information about a paper either from the cover of the journal issue or from the journal's Table of Contents. In both places, information is frequently misrepresented. Even if it is not, mistakes about an article's first or last pages can easily arise. To ensure accuracy, you yourself should gaze upon the actual paper; and you will also be able to say truthfully that you have "looked" at the paper.

Purists also will tell you that each reference should include an **MR** (or *Math. Reviews*) number. Such an addition, often quite convenient for the reader, does demand extra work from the writer. The advent of `MathSciNet` (Section 7.2), the web service available from the American Mathematical Society, has nevertheless facilitated this task. There is now no legitimate reason for omitting this number.

Your bibliography should include only those references that you actually cite in the text. We are frequently tempted to include extra references either for sentimental reasons or because we think that these references might be handy for the reader. The former motivation is spurious, and the latter misguided. If you have no reason to cite a reference, then you also are offering nothing by listing it.[3]

## 2.6.2   Formatting Your Bibliography

There are many possible formats for bibliographic entries. The LaTeX resource BibTeX offers a number of standard bibliographic formats. It is easy to choose whichever one you like.

Of course you can, if you wish, select your own bibliographic style. At the beginning of my career I picked a favorite journal (in fact it was *Mathematische Annalen*) and adopted its bibliographic style. I chose a commonly used format, and it has served me well. Here are two bibliographic references formatted in that style:

> [Bat]   Gill Bates, *How I Made My First Billion*, 2nd ed., Acquisitive Press, New York, 1986.

> [Beh]   Viscount Hugh Behave, Some theories on the gentle art of belching in public, *The Journal of Eminently Forgettable Theories* 42(1976), 35–53.

The first of these is a book, and the second a paper in a journal. Notice that the information provided for a book is different from that provided for a journal article or paper. For a book, the author, title, edition number (if this is not the first edition), publisher, city of publication, and year of publication are usually considered complete bibliographical data. [Some people will include the ISBN.] These are shown in order, and with the correct punctuation, in the example given. For a paper, the author, title, journal, volume number of the journal, year, and pages are usually considered complete bibliographical data. Some people include the issue number of the journal. Of course the protocol for a preprint, for a conference proceedings, for an unpublished manuscript, for a translated paper, and for a Ph.D. thesis are all a bit different. I shall not go into all the details here. The software BibTeX (part of LaTeX) provides particular formats for all these special types of references. See [SG, pp. 407-410] or [Hig] or [VanL] for particulars.

Following the bibliographic style of *Mathematische Annalen*, I always put

- The title of an article in roman font.

- The title of a journal in italic font.

- The title of a book in italic font.

I have become so inured to this paradigm that anything else looks incorrect to me. Others may differ.

---

[3]What some writers (at least in writing a book) do is, in addition to writing a Bibliography, also provide a list called "Additional Reading." This list consists of books, and perhaps papers, that are *not* cited in the text.

### 2.6.3   Different Bibliographic Methodologies

Some people identify each bibliographic reference with an acronym. Not everyone likes the use of acronyms for citing elements of the bibliography. Some people prefer to number the elements of the bibliography from [1] to [n]. The method of enumeration has the disadvantage that, if you add or delete a reference late in the game, it throws off all your numbering. However, using good software can circumvent that problem (see below). Both the numbering scheme and the acronym scheme have the disadvantage that even a one-character typographical error can make it virtually impossible for the reader to tell which reference was intended.

One excellent scheme for bibliographic references, and one virtually essential when the bibliography is long, is illustrated in the following example. It lists three works by John Q. Public, just as they might appear in a bibliography.

John Q. Public

[1987]   *Why I Never Vote*, Ignoramus Press, Brooklyn, NY.
[1992a]  *The Less I Know, the Better*, Rosicrucian Press,
         Poughkeepsie, NY.
[1992b]  On Doctoring Polls, *The Smart Pollster* 31, 59-71.

As you can see, each item by John Q. Public is identified by the year in which it was published. If two items are published in 1992, then the tags are "1992a" and "1992b." If you use this system (known as the *Harvard system*), then when you refer to a bibliographical item in the text you say "By J. Q. Public [1992b], we know that . . . ."

Note that in mathematics we do not usually put bibliographical references in footnotes (however, it *is* customary in certain statistical work, and it was fairly common in mathematics one hundred years ago). This habit came about in part because typesetters objected to the expense and trouble of typesetting copious footnotes. With the advent of TeX, that particular objection is moot. However, the convention persists. In fact, if you were to submit to most mathematics journals a paper with all the references in footnotes, then you would most likely be asked to reformat it. The trouble with using footnotes in a mathematics paper is that the footnote tags can be mistaken for exponents.

If you are writing your paper in LaTeX, then you have the option of using LaTeX's bibliographic utilities. One of LaTeX's features allows you to assign a nickname to each of your bibliographical references. Then, in the text, you can cite any reference by its nickname. When you compile your `*.tex` file, each nickname citation is replaced by the appropriate author-preassigned acronym or machine-preassigned number; the full bibliographic citation occurs at the end of the document as usual (see Section 6.5 for more on TeX and LaTeX).

Slightly more sophisticated is BibTeX's bibliographic database system. With this device, you never write another bibliography as such. You simply have an ever-growing database of bibliographic references. Whenever a new reference comes to hand, you add it directly to the database. There is a very particular syntax for entering each of these references in your database. But it is very easy to learn and to use.

Each reference has an author-preassigned acronym and an author-preassigned nickname. Then, when you are writing a new document, you make a reference by referring to the appropriate nickname in the database (if you cannot remember all the nicknames—perhaps your database has thousands of items in it!—you can just pull the database into a window with your text editor and check it). When you compile the document, a beautiful bibliography is created for you, with the requisite information pulled in from the database.

If the last systems do not appeal to you, then you also can keep the TeX files of all your papers in a single directory. Most of us tend to use many of the same references repeatedly. Thus, when you are writing a new paper and need a reference, you can open a window with your text editor, pull in an earlier paper that has the reference, and cut and paste the reference into your new document.

### 2.6.4   Nicknames

Incidentally, LaTeX also allows you to assign nicknames to your equations and theorems. You can refer to them, during the writing process, by nickname. Then, when the paper compiles, the correct line numbers and theorem numbers are inserted for you automatically. This system is particularly convenient because, if you are in the habit of moving theorems and equations around, you do not have to worry about changing all the numbers. LaTeX keeps track of everything for you.

And now a negative admonition. I know mathematicians—excellent ones—whose bibliographies look like this:

1. Knuth, 1992.

2. Lister, 1991.

3. Machedon, 1988.

This is it! No titles, no journal names, no volume numbers, no page references. This scheme in effect takes the LaTeX device to the limit: you just supply the nicknames but none of the details.

The practice of listing abbreviations in lieu of correct bibliographic references is irresponsible. In truth, such sloppiness should have been caught by the editor, who should have demanded that the author rectify the matter. As indicated at the beginning of this section, the bibliography is part of your paper trail. You hold the responsibility for providing complete bibliographic information. It should be complete in the sense that you have cited everyone who merits citation, but it also should be complete in the sense that all the information is there. The bibliographic sample just provided might mean something to a few experts for a few years. In fifty years it would not mean anything to anyone.

And speaking of "meaning nothing to anyone," do *not* give in-text bibliographic references that have the form "see Dunford and Schwartz" (for those not in the know, [DS] is a three-volume work totaling more than 2500 pages). The only correct and thorough way to give a reference is to cite the specific theorem or the specific page. To be sure, to conserve space and to prevent repetition, we sometimes say "by a variant of theorem thus and such" or "by a variant of the argument in this paper" (the subject of analysis, in particular, seems to be littered with references of this nature). If you find such references necessary in your own work, be as specific as you can so that the reader may follow your path.

## 2.6.5   Styles for Citation

Now let us treat styles for citation. In this section, I have spoken of bibliographic references with the assumption that they will occur on the fly, right in the text. For example:

> By a theorem of Steenrod [Ste], we know that every instance of generalized nonsense is a generalization of specific nonsense.

The good feature of this methodology is that it tells you right away what the source is. The bad feature is that it can clutter up the text a bit. In most mathematics *papers*, the on-the-fly style is used. You make a reference either by acronym, or by number, or by author surname, but the reference occurs at the moment of impact.

In [Ste], Steenrod fulminates against this bibliographic style for the writing of a book. His preference is to have a paragraph or more at the end of each chapter detailing the genesis, development, and sources for the theorems in that chapter. This methodology is commonly known as the "Princeton style."

Many books in the Princeton book series *Annals of Mathematics Studies* handle bibliographic references in this fashion. These little end-of-chapter essays can be quite informative and, if well written, can give the reader a sense of the historical flow of thought that in-context references (as indicated above) do not. I would say that the down side of this end-of-chapter approach is that it serves the big shots well. And it serves the rest of us poorly. If you are annotating a chapter on singular integrals, then you will certainly not overlook Calderón, Zygmund, Stein, and the other major figures. But you might overlook the smaller contributors. The in-text, on-the-fly reference method largely precludes this problem because it systematically holds you accountable: you state a theorem, and you give the reference; you recall an idea, and you give the reference. You are much less likely to give someone short shrift if you adhere to this more pedestrian methodology. Of course the final choice is up to you.

## 2.7   What to Do with the Paper Once It Is Written

### 2.7.1   What to Do with the Paper

Ours is a profession where, by and large, we are left on our own to figure out how to function. Nobody shows us how to teach, nobody tells us how to write a paper, and nobody tells us how to get published. This section addresses the last issue.

So imagine that you have written a paper that you think is good. How do you know it is good? Being a mathematician is a bit like being a manic depressive: you spend your life alternating between giddy elation and black despair. You will have difficulty being objective about your own work: before a problem is solved, it seems to be mightily important; after it is solved, the whole matter seems trivial and you wonder how you could have spent so much time on it. How do you cut through this imbroglio?

If you are smart, you have told some colleagues about your results. Perhaps you have given some seminars about it. You have sent preprints (either by email or by postal mail) to colleagues. If you have kept your ears open, you will now have some sense of how receptive the world is to your ideas. Are your listeners surprised, impressed, confused, bored? Sometimes they will suggest changes. Consider all criticisms and suggestions carefully, and make appropriate changes to your paper. And then decide where to submit it.

But before you make that momentous decision, let us back-pedal a minute and address the question of how to decide when you have something worth writing up. You need to learn how to decide which problems are worth tackling and which results are actually worth writing up and publishing. These are sometimes confusing issues that every mathematician must learn to face.

We all know that the keys to success in this profession of ours include intelligence, perseverance, and hard, tedious work (not necessarily in that order). Some may deny it, but "art" is also sometimes relevant. Let me explain. Ideally, the working mathematician sets a problem for himself/herself: solve the (restricted) Burnside problem, or calculate the dual of the Hardy space $H^1$, or prove the corona theorem in several complex variables. There are extraordinary mathematicians who can actually rise to the challenge of this ideal: E. Zelmanov did the first and C. Fefferman did the second. Nobody has done the third, although many of us have tried. For most of us, this point-and-shoot attempt at greatness rarely succeeds.

Another, more modest approach, is this: become completely immersed in a subject, and then formulate a program. Determine to assume hypotheses $A, B, C$ and endeavor to prove conclusion $X$. Sadly, this *modus operandi* is also only occasionally successful.

Indeed, mathematical success is often arbitrary, unpredictable. In fact what happens in practice is that we try a great many things. Some succeed and some do not. Along the way, hypotheses are constantly being altered and substituted and strengthened; conclusions are redirected or transmogrified or reversed. The theorem that we end up proving is rarely the theorem that we set out to prove. As chaotic as this practice may appear, it is perfectly reasonable. Columbus sought a new passage to Japan and instead found America. Jonas Salk discovered the polio vaccine by accident. Milnor discovered multiple differentiable structures on the 7-sphere because calculations on another problem were not working out as planned.

Always remember that a successful mathematician differs from an also-ran because he/she can take his/her partial results and his/her tries—and yes, even his/her failures[4]—and turn them into an attractive tapestry of theorems and corollaries and partial results and conjectures; the latter instead takes two years of hard work and dumps it in the trash.

As you read these words, do not suppose that I am advocating any degree of chicanery, or self-promotion, or hype. I am instead encouraging you to have the confidence and fortitude to make something of your work. I want you to have confidence in what you are doing, even when it seems to be going nowhere. Part of doing mathematics successfully is to get in there and calculate and reason and think and ponder. But another part is to evaluate and organize and deduce.

What I am describing is a bit easier to imagine for a laboratory scientist. He/she performs a huge experiment that may take a year or two and may cost a few million dollars. No matter how things turn out, he/she must make a show of it. He/she must report to his/her granting agency and write papers about how his/her laboratory has been spending its time and effort. The message here is that you, as a mathematician, must do something similar, but you must survive without so many external stimuli. Not only your ideas but also your motivation must all reside in your head. Part of training yourself to survive in this profession is training yourself to be stubborn, to be proud, to be absolutely determined to succeed.

## 2.7.2 Making Progress

I must note, of course, that another key to success is actually making some progress. It just will not do to tell yourself (and the world) that for the next twenty years you will work on the Riemann hypothesis, *unless you can arrange to have something to show along the way.* You do not get tenure, or a promotion, or an invitation to the International Congress by advertising that you are working on a great problem and telling people that they should contact you a generation later to see how things worked out. I have a friend who has a twenty-five-step program for proving the Riemann hypothesis: "Count to twenty-four and then prove the Riemann hypothesis." There is wisdom in this little joke. The successful mathematician knows how to manage his/her research program so that it proceeds incrementally, so that he/she can report progress along the way—including writing up papers and giving talks and demonstrating that

---

[4]A twentieth-century Hungarian philosopher once said that a mathematician is nothing but a collection of statements that he/she cannot prove.

he/she is making progress towards a larger goal. By the same token, the good mathematician knows how to determine when he/she is *not* making progress, when his/her program is *not* paying off, when it is time to move on to something else. You can always keep that unsuccessful problem in the back of your mind and return to it later (presumably with new tools and new experience to apply to it).

### 2.7.3 Publishing Your Paper

Let us suppose that you have organized some of your material and turned it into a paper. You believe that this is a worthy piece of work. You want to get it published. You need to bring a good dose of wisdom to the submission process.

Keep in mind that the one hard and fast rule in this business is that you can submit a paper to just one journal at a time. *Never consider deviating from this policy.* In the words of Clint Eastwood, "Don't even think it." If you do send the same paper to two different journals simultaneously, then that paper is liable to be sent to the same referee by both journals; thus you will be caught red handed! Agonizing though it may be, you must wait for a decision from journal $n$ before you submit to journal $(n+1)$. As a result, there is considerable motivation to exercise wisdom when choosing a journal.

There is a distinguished mathematician, now retired, who in his heyday wrote about a dozen papers per year. He submitted them all to the *Annals of Mathematics*. Several of his papers were accepted by the *Annals*. Others were either rejected or else the author was asked to perform various revisions. Now, writing twelve papers per annum as he did, this mathematician had no time for revisions. So, in cases two and three, he sent his papers to a well-known journal that was reputed to have minimal standards (what the famous computer scientist Dijkstra would call a "write-only" journal). Thus this esteemed man has a publication list, emblazoned in `MathSciNet` for all to see, consisting of several citations in the *Annals* alternating with citations in this other "catch-all" journal.

Another famous mathematician was in the habit of bringing his latest preprint to the departmental secretary, together with a list of journals to which it might be submitted. Her job was to cycle through the journals on the list, one by one, and to inform the professor when his paper was finally accepted. In this way the good scholar was spared the grief of dealing with surly referees and uncooperative editors.

The preceding two strategies are amusing but probably unwise for most mathematicians. The working mathematician should have a sense of which are the very best journals, which are at the next level, and which are of average quality. How can one gauge which journals are which? They all look rather elegant, and all profess to have high standards. They all have distinguished people on their editorial boards. What is the trick?

Part of the secret to success in this profession is to talk to people. Doing so, you will quickly learn that *Acta Mathematica*, the *Annals*, *Inventiones*, and the *Journal of the American Mathematical Society* are four of the pre-eminent mathematics journals. These four obviously reflect my prejudices as an analyst. Others might name the *Journal of Differential Geometry* or the *Journal of Algebra* or the *Journal of Symbolic Logic* as being at the top. Opinions will vary. Perhaps *Duke*, the *Transactions of the AMS*, the *Journal of Geometric Analysis*, and several others are at the next level. And on it goes. There are prestigious journals and there are excellent journals. Many journals fit into both categories, and many fit into neither. There are nearly 2000 mathematics journals in the world,[5] so you have many choices of where to publish.

### 2.7.4 Which Are the Best Journals?

Begin by considering where cognate results have appeared. The *Journal of Algebra* will probably not consider papers on singular integrals. The *Journal of Symbolic Logic* probably does not publish papers on Gelfand–Fuks cohomology. Certain journals have become the default forum for work on operator theory or several complex variables or potential theory. Consider those if your work fits. Try to predict which editors will understand what your paper is about and will understand your paper well enough to select an appropriate referee. You need not actually *know* the editor, but it is comforting to know what the editor's values are and what mathematics he/she practices and likes.

Every journal has certain standards and mores to which it subscribes when it evaluates your work. Among these are

- Is your paper relevant to the journal's mission?

- Is your paper clearly expressed?

- Is your paper objectively formulated?

- Does your paper suitably acknowledge prior work?

- Is your paper free from plagiarism—both plagiarism of other scholars and plagiarism of yourself?

- What exactly is the question being analyzed here, and how does it fit into the big picture?

- Do you provide sufficient detail of your calculations (in some cases this will include computer calculations) so that others may reproduce them?

- Are the proofs complete? Have the conclusions been cogently drawn?

- Are the results original, significant, and new?

You should really think carefully about all these points as you decide which journal is the right one for your work.

If you submit your work to a journal of the highest rank, then you might pay in several ways: **(1)** the refereeing process may take an extra long time, **(2)**

---

[5]By contrast, geophysics has only about five journals.

the journal may have a huge backlog, **(3)** the paper may be rejected for almost any reason. Thus the entire process of getting your work published could drag on for two years or more. If you are fighting the tenure clock, this could be a problem. In some ways it is better to err on the low side. Usually mathematical work is judged on its own merits. Nobody will downgrade your work, or you, if your theorems are not published in the optimal journal. But do not publish in an obscure journal that nobody ever reads.

You can form your own opinion of journals by seeing what papers they publish and by which authors; you can look at how many truly eminent people (and from which universities) are on their editorial boards, and you can learn something just by submitting your papers to various journals and seeing what happens.

Since the latter strategy is costly—in terms of time, and perhaps your bruised feelings as well—you should develop a sense of what is a typical *Annals* paper, what is a typical *Transactions* paper, and what is a typical *Rocky Mountain Journal* paper. If you are in doubt, ask someone with more experience. If someone whom you respect and trust has read your preprint, then he/she would be an ideal person to ask for suggestions as to where to submit.

### 2.7.5   Journal Backlog

Consider as well the backlog of the publication, the turnaround time from submission to acceptance (or rejection), and similar data. Fortunately, the *Notices of the AMS* publishes, at least once per year, a detailed analysis—containing just this sort of information—of all the major journals. A year may pass while you are waiting for an acceptance or rejection, and your tenure clock may be ticking. So you need to choose wisely.

Of course you will learn from experience. You also will have to decide for yourself whether to shoot high and take your chances, or to shoot low and optimize your likelihood of a quick acceptance. If your tenure case is a few years down the road, then this choice should not be taken lightly. Deans tend to know which are the good journals and which are not. (In fact, I know of several universities where the dean has circulated a ranked list of mathematics journals. The implication is that "If you want to get promoted, then you had better publish in these journals but not in those journals.") In particular, deans are not impressed by a young assistant professor whose work is all submitted to "gimme" journals. They are also not impressed by a dossier with most papers "submitted" but not yet accepted.

You need to give some thought to the format of your paper, because you want your work to appear professional. Most journals have a section or a web page called "Instructions to Authors" or "Instructions for Submission." Before you submit paper $X$ to journal $A$, you should read those instructions. They will tell you how many copies are needed, whether the title and abstract and other data should be on a separate page, whether the journal requires key words and AMS Subject Classification Numbers, what languages the journal will accept (English, French, and German are the most common—though there *are* mathematics journals that will take papers in Latin or Esperanto or Japanese or Spanish or Russian), any formatting requirements, length restrictions and where to send the paper (to the Editorial Office, or to an Associate Editor of your choosing, or perhaps another option). You need to know whether the journal prefers submissions in TeX, whether the journal has a TeX style file that you should use, whether the journal accepts electronic submissions, and so forth.[6] You will annoy the editors, and cause unnecessary delays and confusion, if you do not follow these readily available instructions. Of course, many journals now take electronic submissions (i.e., uploading at an Internet site), and this system *forces* you to make the right choices. Some journals just ask you to send the paper (in `*.pdf` format) as an email attachment to one of the editors.

## 2.7.6 Web Submission

It must be noted that, in today's world, it is common to submit a paper to a journal by way of a website. You simply fill out some web forms and upload your paper and you are finished. With this system, you certainly do not need to worry about how many copies to submit. And you will be prompted for all the ingredients that are needed. It is a reliable system, and it works.

The journal will assume that the "communicating author" is the person who submitted the paper—unless you explicitly tell it otherwise. All further correspondence will be conducted with that person at that email address.

Even if you submit your paper at a website, you will often be asked to supply a cover letter. At the website this will be in electronic form. Some authors think that the cover letter is an opportunity to make a pitch for the paper. Such an author will fill the cover letter with fulsome praise of what is in the preprint, why it improves on the existing literature, and who might be a suitable referee. Most editors will not find such remarks helpful, and many will find them annoying. By naming potential referees, you may in fact be ruling them out in the mind of the editor (since he/she may think that they are your pals). In short, keep the cover letter simple and dispassionate.

After you have submitted your paper to a journal (allowing perhaps for a short delay), the journal will notify you that it has received the paper; this is usually done by email. The journal will often assign a manuscript number to your paper and will advise you to use this number in all future correspondence. I run a journal, and I can tell you that this number is valuable. The journal office can easily misfile a paper with multiple authors; also, since the paper is passed from Managing Editor to Associate Editor to one or more referees, the paper can be misplaced. It helps significantly when authors and editors use the manuscript number. Such a number might be "JGEA-D-16-243," indicating that this is the 243rd paper received in 2016 by the *Journal of Geometric Analysis*. The

---

[6]I know quite a lot about TeX. Indeed, I have written a book about it—see [SaK]. But I find it difficult to learn somebody else's TeX style files on the fly. I can tell you from my experience that most journals have a TeX style file, and most will *not* insist that you use it. It tends to be easier for everyone if you just create your work in out-of-the-box LaTeX and let the journal convert it to their local TeX style.

notification email from the journal will conclude by saying something like "Don't call us; we'll call you." In other words, you may have to wait a while for the referee's report; so sit tight.

### 2.7.7  Waiting for a Decision

Expect to wait four to six months for a report. After that wait, you are well within your rights to send a polite note to the editor to whom you submitted the paper;[7] simply state that you submitted the paper on thus and such a date, received an acknowledgment on another date, and you are now wondering whether there has been any progress in the matter. Most editors appreciate a gentle reminder, and will in turn nudge the appropriate Associate Editor or referee.

Eventually you will receive a referee's report. It may be a paragraph or it may be five pages or more. It may say "This paper is terrific. Publish it as quickly as you can." Or it may say "This paper is dreadful. Stay as far away from it as you can." Most often it will say something in between these extremes.

If the paper is rejected, then you will have to ply your wares elsewhere. A rejection does not necessarily mean that the paper is bad or that its results have no value. Many journals suffer from a serious backlog, and thus send most papers back unread (this is, properly speaking, not a rejection—for the paper has not even been examined or evaluated); sometimes the editor picks the wrong referee, or a referee with an ax to grind, or a referee who did not understand the paper; sometimes the editor misunderstands the referee's report; sometimes the referee is just plain wrong. Some of my own most influential papers have been treated rather shabbily. I know even Fields Medalists who tell horror stories of papers rejected. As I have already said, one secret to success in the academic game is perseverance. If your paper is accepted the first time around, then congratulations. If not, then you should try to be objective and figure out why. Then act intelligently on that new information.

One point that needs to be made clearly is this: If your paper is rejected, then *do not* plan on revising it and then resubmitting it to that same journal. If a journal rejects your paper, then you should assume that that journal does not want your paper. You should submit your work elsewhere.

### 2.7.8  Bringing the Matter to Conclusion

If your paper is accepted, then the referee will most likely have offered comments and suggestions. Some referees go so far as to suggest alternative proofs, different references, or entirely different approaches. Some editors will instruct you to read the referee's remarks, make those changes that you wish, and then submit the final version of the manuscript, labeled "revised" and with a new submission date, to the journal; other editors will explicitly make final acceptance conditional on your responding in detail to everything that the referee

---

[7]This is the stage where it is important that you have the editor's name, his/her email address, and the manuscript number at hand.

has said. In this last case, if you want to continue doing business with the journal (you always have the fallback option of withdrawing the paper), then you are honor bound to respond to *each of the referee's remarks*. The best way to respond is to treat the referee's remarks one by one, and to record in a cover letter to the editor a brief description of just what you did in each instance. In some cases, you may say "the referee is mistaken and here is why." Or you could even say "this is a matter of taste and I respectfully disagree." In most instances, though, you can expect the referee's comments to be accurate and useful, and you will probably want to implement them in some form.

If you feel that the referee has been particularly helpful, then you may wish to add a sentence to the paper—alongside your other acknowledgments—saying that you thank the referee for useful suggestions. Refereeing is generally done anonymously, so do not plan to mention the referee by name.

When you are finished with the requested revisions, then make the usual number of copies of the revised manuscript, mark each of them "Revised" and date them, and then return these to the editor along with a new cover letter. Your new cover letter should state plainly that this is a revision of a previously submitted paper, that you have responded to the referee's remarks, and that you consider this new version to be the final copy.[8] Please note, however, that the editor *might* send the revised paper to the original referee—or to some other referee—again, and you may be asked to make even further changes. You can expect to receive an acknowledgment of your new submission, together with a clear statement of whether this is the end of the road or whether you will be hearing again from a referee.

### 2.7.9 Skillful Use of T<sub>E</sub>X

And now, as the Managing Editor of a journal, I would like to ask you a favor. It is quite common for a young mathematician to learn TeX on the local computer system at his/her University $X$. That system will no doubt have in-house macros and fonts that everyone at University $X$ uses. The trouble with this is that the TeX file for the paper will compile on the computer at *that* university, but not on other computers.

You must learn to make your papers self-contained, so that all the macros and font calls are in the source code file for the paper. Usually, when you first submit a paper electronically, you will only send in a `*.pdf` file. Since the referee and/or the editor will likely ask for revisions, there is no sense in sending the TeX source code file at that time. But, after the paper is accepted and in final form, then you *will* submit your TeX source code file. And you will want the journal to be able to compile your file. So pay attention to the points made in this and the preceding paragraph. See also the discussion of TeX in Chapter 6.

---

[8]Of course if you are dealing with the journal by way of a website, then the procedure is different. You will submit your revision by uploading it to a suitable node. You will also fill out various forms telling the journal just what you are doing. The editors will automatically be informed of receipt of your revision, and the process will move ahead.

### 2.7.10   Galley Proofs

After a suitable number of iterations of the procedures just described, you and
the journal will reach some closure. Then you must wait—this wait could be
from one to several months or more—for the galley proofs or page proofs of
your paper.[9] These you must proofread meticulously, both for mathematical
accuracy and for typesetting accuracy. There also will be "Author Queries,"
noted by hand, on the proof sheets. You must respond to each of these. Often
you will handle these matters by email. Sometimes you will instead be sent a
`*.pdf` file which you will annotate with "electronic sticky notes."

You will always be asked to turn your proof sheets around rather quickly—
often within 48 hours. You will sometimes be asked to sign a statement saying
that you approve this version of the manuscript going into print.

It is quite standard at this stage for the journal to ask you to sign a copyright
transfer agreement. As noted elsewhere in this book, when you write something
it is immediately copyrighted *to you*. But the publisher of the journal in which
your paper is to appear will typically want to hold the copyright on your paper.
Some authors object to this because of the possibility that, years later, someone
will want to put the paper in a volume of collected work on a certain topic. If
the paper is copyrighted to a different publisher, this will give rise to unpleasant
negotiations and nasty fees. Fortunately, most journal publishers are willing to
talk to you about who holds the copyright.

*You will also be asked at this time how many reprints you want.* In truth,
reprints are these days something of an atavism. Typically a journal will give
you a `*.pdf` file of the final version of your article, so that you can then print all
the reprints that you may want. In any case, after you send back your response
to the galley proofs, your job is done. Just wait for your paper to show up in
the library, and you will know that your paper is now part of the permanent
archive.

## 2.8   The Importance of Notation

### 2.8.1   The Use of Notation

One thing that makes mathematical writing special is our use of notation. It
is just a fact that many mathematical facts and assertions are best understood
*not* by expressing them in words but rather by formulating them in suitable
notation. In fact an examination of the history of mathematics in the Middle

---

[9]It should be noted, however, that some journals now instantly publish your paper on
the web as soon as it is accepted. Most authors are pleased at this eventuality. But it is
understood that the archival copy of the paper is the hard copy version, the one that has
undergone copy editing and other vetting.

Since the hard copy version of the journal is published in a queue, there may be a
considerable delay before the printed version of your paper appears. It is also the case these
days that some journals do no copy editing whatsoever. Whatever the liabilities of such a
policy, it certainly streamlines the process.

Ages and the Renaissance shows clearly that the advance of the subject was held back considerably by lack of notation. See [Caj] for a fascinating history of mathematical notation.

Much mathematical notation is standardized, and of course you should use that standard notation to make your thoughts as clear and widely accessible as possible. Everyone denotes an integral by $\int$ and a sum by $\sum$ and a derivative by $d/dx$, and you should too. But there will be times when you need to introduce your own notation for some new idea or concept that you have formulated. You should think carefully about how to do this. You want your new notation to have the following properties:

- It should be as simple as possible.

- It should be easy to write and easy to read.

- It should not conflict with other, well-established and widely used, notation.

- It should convey your thought clearly.

- It should be consistent and not easily confused with other notation in your paper or book.

### 2.8.2 New Notation

It is probably a good idea to try out any new notation on some graduate students or some colleagues. Give a seminar in which you trot out the notation and see how people react to it.

It is quite common, when you write an elementary textbook, for the publisher to hire a third party to prepare the solutions manuals. All well and good. But do be sure that this third party follows your notational conventions so that the solutions manuals are consistent with the text. Too many textbooks have failed miserably because this simple precaution was not followed.

Finally, once you have introduced a new notation, then you should try to stick with it. If you write a paper on subject $X$ in year $N$, then your subsequent paper on that same subject in year $N + k$ should really, if at all possible, follow the same notational conventions. This is another reason for you to be extra careful in selecting a new notation to begin with.

## 2.9 A Coda on Collaborative Work

### 2.9.1 Choosing Your Collaborators

I have written a great many collaborative papers, and some collaborative books as well. I know others who have never collaborated. And there are others still who have collaborated a few times and would never do so again. Which characteristics lead to a successful and happy collaboration and which do not?

First, if you agree to collaborate on a project (and both parties had better agree at the outset; do not leave this question until the project is finished!), then set aside all questions of priorities. At the end of the collaborative process, it is both painful and inappropriate for one author to say "Well, you didn't contribute very much. My name should go first" or, worse, "Your name should not appear at all." If, at the end of the first paper, either or both participants deem the collaboration unsatisfactory, then the authors can go their separate ways. But, in my view, an agreement to collaborate is an *a priori* contract to see things through to the end.

Some of my collaborations involve multiple papers; in one case the joint work amounts to thirteen papers and two books. Another collaboration of mine involves six papers and four books. In these monumental collaborations, both my collaborator and I know that on some papers he/she contributed more and on others I contributed more. It is the same for the books. I can honestly say that neither of us dwells on the matter. Taken as a whole, we are both quite pleased and proud of the *oeuvre*. In the first of these collaborations, one of us has lost interest in this subject area and the other one has pushed on a bit further, either writing papers alone or in collaboration with others. This has worked out well, because each of us respects the other.

## 2.9.2   Respecting Your Collaborator

And this last point is the real key. I know of collaborations in which one author purposely introduced errors into the joint paper in order to see whether the other author was truly reading the paper or not. I know of a collaboration that got to the stage of the paper being submitted to a journal; after a time the authors had a dispute, and one author unilaterally withdrew the paper and resubmitted it elsewhere under his name alone. I know of a collaboration between two lifelong friends who were developing their twentieth paper together; in this latest paper, they could not agree on whether to call the first result Theorem 1A or Theorem A1; the matter ended with lawyers, death threats, and guns brandished in the air (this *really* happened). In all these cases the base of the difficulties was that the authors did not respect each other.

None of the scenarios described in the last paragraph should take place when you engage in a collaborative effort. You should view a collaboration as an adventure in which each of you will see what you can derive from it. Each of you should respect the other(s), and take great pains to be courteous and helpful to all concerned. The goal is to produce a nice piece of work—*not* to squabble over credit, *not* to argue over whose name should go first (alphabetical is almost always best), *not* to argue over whether future papers will be joint or will be written separately.

## 2.9.3   Sticky Wickets

Unfortunately, this last point *can* lead to sticky wickets, even between the most well meaning of participants. Imagine this scenario: Mathematicians *A* and

$B$ write one or more joint papers. The collaboration then seems to go into remission. Each author goes his/her own way, and they have little contact for a couple of years. Then one of the authors (say $A$) cooks up another idea and writes a new paper, by himself, which in some sense builds on the ideas in the old series of joint papers. Mathematician $B$ gets wind of this new paper, feels that his contribution to the earlier work justifies his name appearing on this new paper, and relates this feeling to $A$ in no uncertain terms. Mathematician $A$ feels that the joint work was long ago and far away. The main reason for the existence of the new paper is *his* new idea—which is due to him alone. Mathematician $A$ feels that $B$ has already received adequate credit for the joint work; no further credit is due $B$. As you can imagine, a major fight ensues.

This situation is most unfortunate. Nobody is right and nobody is wrong. Here is what *should* have happened—in the best of all possible worlds. Realizing that the new paper builds on old joint work with $B$, mathematician $A$ should have phoned $B$ and told him about it and then said "I think it would be appropriate for this new paper to be joint between us. What do you think?" Mathematician $B$, ever the gentleman, should then have said "Oh no, this is your idea. Write the paper by yourself. You can thank me in the introduction if you like." Having participated in transactions of this nature, I can tell you that this courtesy is a most satisfactory way to handle the matter. Typically, mathematician $B$ is not hungry for another paper; he/she just wants his/her due. Typically, mathematician $A$ is not anxious to offend $B$; he/she just wants credit for his/her new idea. (Of course the human condition is such that there are always more complex forces at play. Perhaps $A$ feels that, in the world at large, $B$ is generally given more credit for the collaborative work than $A$. Perhaps $B$ feels that $A$ never pulled his weight in the first place and therefore $A$ owes $B$. Fortunately, this is not a tract on psychology, so I shall not comment further on these complexities.)

By touching base with your collaborators in a respectful fashion, you can usually avoid friction. And the effort is worth it. To go to a conference and run into a former collaborator is a pleasure. If the relationship is healthy and friendly, then there is plenty to discuss, and the potential for future joint work always lies in the offing. If instead there is friction and resentment between you and your former collaborator, then meeting again can be perfectly dreadful. Bend over backwards to avoid such a liability.

## 2.10 Recognition of Good Writing

Much of the reward for good writing is just the personal satisfaction of doing a good job. But it is always good to hear from your readers, to know that you are actually reaching someone, to understand that your words are resonating with the world out there.

Especially gratifying is to win a writing prize. Today there are a good many of these. Among the most prestigious writing prizes are the three Steele Prizes of the American Mathematical Society and the Chauvenet Prize of the

Mathematical Association of America. There is also the Lester R. Ford Award of the MAA, the Euler Book Prize of the MAA, the AMS Book Award, the Edwin Beckenbach Prize of the MAA, and the Levi L. Conant Prize of the AMS. It is quite prestigious, and certainly a pleasure, to be awarded one of these prizes.

## 2.11 Closing Thoughts on Effective Mathematical Writing

As Ludwig Mies van der Rohe said, God is in the details. Good writing consists of attending to a great many particulars, from grammar and syntax to logic and organization.

It is our hope that this chapter has helped to orient you towards the process. Now the ball is in your court. You must apply yourself to the art of writing, develop your skills, and communicate with the world.

# Chapter 3

# Exposition

*Reading maketh a full man, conference a ready man, and writing an exact man.*

Francis Bacon
*Essays* [1625], Of Studies

*You can fool all of the people all of the time if the advertising is right and the budget is big enough.*

Joseph E. Levine

*When Kissinger can get the Nobel Peace Prize, what is there left for satire?*

Tom Lehrer

*Life should be as simple as possible, but not one bit simpler.*

ascribed to Albert Einstein

*If you have one strong idea, you can't help repeating it and embroidering it. Sometimes I think that authors should write one novel and then be put in a gas chamber.*

John P. Marquand

*Writing comes more easily if you have something to say.*

Scholem Asch

*Considering the multitude of mortals that handle the pen in these days, and can mostly spell, and write without glaring violations of grammar, the question naturally arises: How is it, then, that no work proceeds from them, bearing any stamp of authenticity and permanence; of worth for more than one day?*

Thomas Carlyle
*Biography* (1832)

## 3.1   What Is Exposition?

### 3.1.1   Surveys and Other Exposition

Perhaps the highest and purest form of mathematical writing is the research paper. A research paper, in its best incarnation, contributes something useful and insightful to our collective mathematical knowledge. If it is very good, then the contribution may live for a long time. The creation and publication of research is what mathematics is all about.

But our profession involves us in other types of writing. We must write letters of recommendation. We must write referee's reports. We must review cases for tenure and promotion. We write surveys. We sometimes write book reviews. We may be called on to write opinion pieces. The present chapter concentrates on such *expository writing*.[1]

In its simplest form, mathematical exposition could take the form of a survey of a field on which you are an expert. Or it could be a text or monograph on some specific area of mathematics. The new challenges present in such a writing task are these: **(i)** you are attempting to reach a broader audience than that which would read one of your research papers; **(ii)** you must strike a balance between how much mathematical detail to give and how much explanation and/or handwaving to provide; **(iii)** you must be open to the idea that this is a new type of writing with new goals and new audiences.

## 3.1.2   The Reader of Exposition

The reader of an expository article does not want to work as hard as the reader of a research article. Envision your reader sitting on a park bench reading your expository article, or putting his/her feet up and drinking a cup of coffee while reading. *Do not* imagine your reader with a pencil gripped in his/her fist, slaving away over each detail of your paper. Thus, if you are writing an article about the influence of the Atiyah-Singer Index Theorem on modern mathematics, you certainly will not prove the theorem. To be sure, you will refer to some of the excellent books on the subject. You will explain how the result is a far-reaching generalization of de Rham's theorem and the Riemann-Roch theorem. You will describe the ingredients of the proof, and will give a rough sketch of its structure. But you will certainly not *prove* the theorem.

You also will not assume that your reader already knows all the jargon in the subject. You will not assume the reader to be expert in $K$-theory or pseudodifferential operators. Nor will you assume that your reader is familiar with the motivation for, and the applications of, the subject. *You should not assume that your reader has the perfect background to read what you are writing.*

So you have your work cut out for you. Expository writing is a lot like teaching. You frequently must anticipate your audience's shortcomings and make suitable adjustments in your presentation. But in expository writing you must be smarter than you are when you are a calculus teacher. In the latter situation, your audience is before you and is sending you signals. When you are writing, your audience is (if you are lucky) only in your head.

---

[1]We save a discussion of book writing for Chapter 5.

## 3.2 How to Write an Expository Article

### 3.2.1 Authoring Exposition

For the purposes of this section, the phrase "expository article" means a survey article. Such an article might be a survey of some field on which you are expert. Perhaps you are one of the pre-eminent experts, and therefore the canonical person to be writing this survey. Or your colleagues have called upon you to plant a flag for the subject. One scenario is that you have been invited to give a one-hour talk at a national meeting of the American Mathematical Society. It is natural, and commonplace, for you to turn such a talk into a survey that will appear in the *Bulletin of the American Mathematical Society.*

But such an august occasion is not a necessary condition for the writing of a survey. You may simply feel that a survey is needed, that certain wrongs need to be righted, or that the time is ripe. More and more journals are soliciting expository articles; there is a market for high-level exposition done well. The journal *Expositiones Mathematicae* specializes in expository articles and publishes substantial articles written at a sophisticated level.

In order to write a good survey article you will need a detailed outline before you. You will be covering a lot of ground, and you do not have the luxury of hiding behind the details of the proofs. In fact, you will most likely not be presenting any proofs in their entirety. When you do choose to present a proof, it will probably be a sketch, or a pseudo-proof. If you are clever, you may present a well-chosen example, work it through, and then say "this reasoning also shows you how the proof works." Or you might say

> This example is in fact the enemy. The proof shows that this example represents the only thing that could possibly go wrong, and then systematically shows that the hypotheses rule the example out.

### 3.2.2 Elements of a Good Survey

A good survey should build to a climax: Poincaré looked at this example, Lefschetz looked at that example, eventually people realized what was going on, and the Eilenberg-Steenrod axioms were formulated (I am thinking here of the genesis of algebraic topology). Alternatively: first there was the Laplacian, then there was general elliptic theory, then there was the $\overline{\partial}$-Neumann problem, then pseudodifferential operators evolved. Simply to begin citing technical results in chronological sequence is not to write an effective survey. You are telling a story, and you must create a tapestry.

A good survey should have a stirring conclusion. By this I do not mean "That's all, folks!!" or a hearty cry for more and better research on Moufang loops. Instead, your survey should conclude by taking a look back at what ground has been covered and where the subject might go in the future. It should note the historical turning points (which you have, I hope, described in the body of your survey), and make speculations about what the future milestones might be. If appropriate, it should sum up what this subject has taught us so far and what it might show us in the years ahead.

A good survey should have an extensive bibliography. You are not doing your job if you merely say "The three standard books in the subject are these, and you should look in their bibliographies for all the technical references." By all means mention the three standard treatises, and extol the virtues of their bibliographies. But you *must* create your own bibliography. Your list of references is your detailed definition of what the subject is, what are the most important papers, and what is the latest hot stuff. Compiling a good bibliography is a lot of work (though, with the aid of modern technology, not nearly so arduous as in years past—see Sections 2.6, 5.5). But this effort is a necessary part of the process, and the result will be a valuable tool for both you and your colleagues in the years to come.

### 3.2.3   Cautions Connected to a Survey

Writing a good survey—one to which people will refer for many years—is one of the hardest writing tasks there is. Getting all the basic ideas on paper, and in the proper order, is just the first step. Once that task is completed, then you must craft the piece into a compelling tale with introduction, entanglements, climax, denouement, and finale (much as in a Shakespearean tragedy).

*Be absolutely certain that you have not slighted any of the players in the subject, nor inadvertently misrepresented their contributions.* Almost certainly, in the course of writing your survey, you will be saying (perhaps *sotto voce*) "Here is the right way to see things (implying, perhaps, that some others are not the right way) and here are the important contributions." Whether you do this consciously or not, you certainly will do it. Take extra care that you are diplomatic, and that you let everyone's voice be heard.

It is generally a good idea, when you write a survey, to send it around to all the key players in the field *before publication*. Give them a chance to respond to it and offer ideas. Most scholars are pleased to be given such an opportunity.

There was something of a mathematical mini-crisis a few years ago when an important mathematician wrote a survey of a topic in harmonic analysis. This was a lovely piece, and I certainly enjoyed reading it. But he made the mistake of *not* showing it to people before it appeared in the *Notices of the AMS*. In those days the *Notices* had a policy that no article could have more than 10 references. Well, you can certainly imagine that a survey will have many more than 10 references. So, when people saw the published article, they were alarmed and offended.

What the author *should* have done is this: First, he should have circulated the article *before* publication. Second, he should have put a line in the article saying that he was restricted to just 10 references, but that a complete list of references appears on thus and such a website. That website could also contain ancillary examples, figures, and other material that would enhance the article. This *modus operandi* would have solved the problem and left everyone happy.

### 3.2.4 The Survey as an Entrée

The reader of your survey should come away from it feeling that he/she has been given an entrée to some new mathematics. He/she should have **(i)** learned some new, and **(ii)** seen some new techniques, and **(iii)** learned some interesting history. Forty pages of descriptive prose, without any substance, will not wash with a mathematical audience. You must sketch how the ideas unfold, and endeavor to give some indications of the proofs. When deciding what to prove, you must balance what is instructive against what is feasible in a short space. Often you can prove only a special case, or you might say "to simplify matters, we add some hypotheses." As an instance, proving the inverse function theorem for a $C^1$ function is hard. But if you assume that the function is $C^2$, then you can use the remainder in the Taylor expansion to good effect, and the proof suddenly becomes easy (see the details in [Kr1]). The key ideas are still present, and they come out much more clearly.

Just as when giving a talk, you can fudge a bit in your survey. State theorems precisely and correctly but, when presenting the proof, say "For simplicity, we consider only a special case" or "For a quick proof, let us assume that the function is actually real analytic." Readers will appreciate being given a nugget of knowledge, without the gory details. A research paper should contain complete proofs—proofs that are categorical and leave no doubt of their correctness. A survey acts as a pointer to the research literature; it is not usually the final word on the proofs.

## 3.3 How to Write an Opinion Piece

### 3.3.1 Formulating Your Opinion

Most mathematicians agree that writing good exposition is considerably more difficult than writing good mathematics. As has already been described in this book, the latter activity makes few demands on your abilities as a creative writer. You need only exercise some taste in organizing and presenting the ideas. However, when you are expositing, then you are less engaged in statement and proof and more engaged in description, explanation, and opinion formation. There is much more latitude, and therefore you, as a writer, must exercise more control.

Let us turn now to the writing of an opinion piece. The writing of an effective opinion piece will involve all the skills already noted as mandatory for good expository writing. But the opinion piece must also, if it is to be effective, have fire and life and drive. It must capture the reader's attention, and it must *convince* him/her of something. How does one go about pulling this off?

A parody of midwestern political oratory has the would-be congressman declaiming

> Agriculture is important.
> Our rivers are full of fish.

The future lies ahead.

I hope that, when you write your opinion piece, your thoughts have more focus than this politician's, and your message is more incisive and more substantive.

First, to repeat one of the main themes of this book, if you are going to write an essay expressing an opinion, then you must have something to say. And you must know clearly and consciously what that something is, and how you propose to formulate it and to defend it. It does not wash to say, in your mind's voice, "I am going to write an essay in opposition to the teaching of calculus in large lectures because it is a bad idea and I hate it." In point of fact most of us agree with this thesis, but it turns out to be extremely difficult to harness facts and arguments to support the thesis. Couple this lack with the fact that there are articulate and serious people—who spend their professional lives studying such matters—who disagree vehemently with the thesis (see [Dub]) and can marshal forceful arguments against it, and your obvious essay in support of an obvious contention suddenly becomes quite painful.

In fact the statistics that bear on the "large lecture" question are a bit unsettling. They tend to suggest that students taught in small classes feel better about themselves and about the subject matter (than do students taught in large classes); they do *not* tend to suggest that such students will turn in a better performance. This is the trouble with facts: they sometimes force you to conclusions that differ with your intuition.

### 3.3.2   Research for an Opinion Piece

What I am suggesting in the preceding paragraphs is that the writing of a position paper or an opinion piece often involves considerable research. This is not the same sort of research that one performs in order to prove the Riemann hypothesis. But it needs to be done, and done thoroughly. There is no substitute for knowing what you are talking about.

You must have a deliberate and explicit formulation of your thesis and your contentions. Best is to enunciate that thesis in your first paragraph. The thesis could constitute your first sentence, or it could be the culmination of suitable background palaver that lays the history and orients the reader's mind toward the main point of your essay. But the thesis you are defending should be put forth—so that it cannot be mistaken—at the outset of your opinion piece.

### 3.3.3   Supporting Your Opinion

The next (major) portion of your essay should consist of cogent presentation of material gathered in support of the previously enunciated thesis. This prose could include facts, reports of studies, anecdotes, analogies, logical arguments, and other materials as well. Note that an essay consisting entirely of anecdotes is at first entertaining but, in the end, not convincing. On the other hand, an essay consisting only of dry facts and logical arguments does not generally hold the reader's attention and is not forceful. (The validity of this last statement depends, of course, on your audience, on your subject matter, and on the context. Obviously, most any mathematical research paper contains just facts and logical arguments. But such a recondite exercise is directed toward specialized researchers with an *a priori* interest in what the writer has to say. The audience for an expository paper or opinion piece is more diverse, less well prepared, and less patient.)

### 3.3.4 Conclusion of an Opinion Piece

The last portion of your position paper should sum up the major points you have made, repeat the most important ideas, and force the desired conclusion. These are the final thoughts with which you will leave the reader. They are analogous to the closing arguments in a jury trial. Weigh each word carefully. Remind the reader what he/she has read, and why.

I do not mean to suggest that persuasive writing is formulaic. This activity is not like laying bricks. Some of the best position papers conform only loosely, if at all, to the rubric just laid out. But the points I have made, and the issues I have raised, are salient to any polemic, no matter what its exact form.

## 3.4 The Spirit of the Preface

### 3.4.1 Why a Preface?

Many writers spend little time in writing a preface; some forget to write it at all. This is a mistake. Your prefatory remarks are often the most important part of your writing. They tell the reader why you write what you write, what your goals are, and what you intend to accomplish. They state what you assume, and what you conclude. Among other things, the Preface will make clear (if you are writing a book) who is the intended audience for the book and what are the prerequisites for reading the book. These principles apply whether you are writing a book (which has a separate, formal preface) or an article (which may have a prefatory section, or collection of paragraphs) or a letter (which may have just one prefatory paragraph). The preface is your statement of purpose. It is vital to your mission.

When an editor at a publishing house receives a mathematical manuscript (for a book, say) for consideration, he/she usually seeks advice from one or more experts in the field. When I am asked by a publisher to review such a manuscript, the first things I look for are a Preface or Prospectus (a marketing version of the Preface) and a Table of Contents. These two items, if present, will give me a quick overview of the project: What material is covered? At what level? Who constitutes the intended audience? What are the prerequisites? What need does this book fill? What are the book's competitors? (In modified form, these queries also apply to an expository article. If I receive a 100-page expository article to review, then I hope that it—or at least the cover letter that accompanies it— contains the information that a book Preface and Table of Contents usually provide.)

Without this information, I have no idea what I am reading. The manuscript could start out with sophomore-level differential equations, and before long be doing canonical transformations for Fourier integral operators. As a result, I have no idea what the author is trying to accomplish.

### 3.4.2  Questions that the Preface Answers

Whether you are writing a research announcement, a research paper, an expository paper, a book, or virtually anything of a scholarly nature, you should always ask yourself the questions in the second paragraph in this section. Most importantly, you must decide *in advance* the book's intended audience and you must, at all times, keep that group clearly in focus. If you are writing a calculus book, then presumably the audience is freshmen, and therefore you must resist the temptation to indulge in asides to the professor. If you are writing a research article, then presumably the audience is fellow researchers, like yourself—not Gauss and God. If I may be permitted a little hyperbole, I will say now that having a strong sense of your audience is the single most important attribute of an effective writer.

### 3.4.3  The Table of Contents

I have sung the praises of the Preface. But the Table of Contents (known as the TOC in the publishing industry) is nearly as important. Here is why. When you are writing a book, which is a big project, you should have the entire scope of your endeavor firmly planted before your mind's eye. In this way you can measure your task, you can see what progress you are making, and you can keep the affair in perspective. Someone who is completely absorbed in his/her mathematics can easily be seduced by each new preprint that comes across his/her desk. This might cause the person to rethink the book every other day. Do *not* fall into that trap. Plan out your book ahead of time and try to stick to that plan. You can put all the new stuff in the next edition.

Sometimes you can have fun just sitting down and starting to write, or just seeing where your thoughts will lead you, or modifying your project every time an interesting new preprint comes across your desk. But let me assure you that these methods are a sure way to guarantee that your book will never be completed. Writing the TOC addresses this impasse.

### 3.4.4  An Onerous Task

Now you may not be interested in writing something like a book. Such a writing project is awesome and onerous; the task is not for everyone. But the principles in the last paragraph apply even to writing a twenty-five-page research paper. You need to have the full scope of the paper in your mind so that you can endow your working methods with a pace and give yourself a sense of incremental accomplishment. This sort of organization is also just a simple device for keeping yourself from becoming completely confused.

I often begin a book by writing the Preface, because it helps me to organize my thoughts and to orient myself toward the project. I refer to it frequently as my work on the project progresses. At the same time, it also makes sense to write the Preface last: for when the book is complete, then you know in detail what you have written and you can describe it lovingly to the reader. My recommendation is to do both. Write a version of the Preface before you begin the book. When you are finished, write it again.

This section on the Preface may seem like a digression, but it is not. Even if you are writing an opinion piece (Section 3.3), or a letter of recommendation (Section 4.1), or a book review (Section 4.2), your piece should contain prefatory remarks. Such remarks are good for your reader, so he/she knows what this piece of writing is supposed to be about. But, most importantly, they are good for you: they keep you honest, and keep you on your course.

## 3.5 How Important Is Exposition?

### 3.5.1 The Significance of Exposition

There are those who will argue that mathematical exposition is not important at all. The one true pursuit, whose fruits are recognizable and of lasting value, is mathematical research. Some would go so far as to claim that even the writing and publishing of research results is an activity suitable only for hacks. Instead, for a good mathematician, it suffices to prove the theorems, tell at least one other person about them, and then let the word spread.

I happen to think that the attitudes described in the preceding paragraph are counterproductive. Scholars are not monks. They are an active and engaged component of society. Along with the universities and research institutes, they are the vessels in which mankind stores its accumulated knowledge and civilization. Thus scholars must communicate. They can do so by giving lectures; the importance of lectures cannot be overemphasized. But scholars also must write. The written word—unlike the spoken word—lasts for ages, and can influence generations.

Exposition is important because it reaches a broader audience than do specialized research articles. Thus good expository articles disseminate information quickly, and they are much more likely to spawn collaboration between different fields than are specialized articles. An outstanding expository article will cause even the experts to reorganize the subject in their own minds.

### 3.5.2 Exposition as Teacher

In my own work, I have found that expository writing is a device for teaching myself. It forces me to organize my thoughts, and to be sure that I understand how a subject is constructed—from the ground up. This is also a device for teaching my students: after I have explained the same idea several times to several different students (perhaps over a period of years), I find it useful to write something down. Then, when the next student comes along, I can give him/her something to keep.

I can still remember, many years ago, reading an article by Freeman Dyson called "Missed Opportunities" [Dys]. I have never seen anything like it before or since. In this article, the author makes statements like the following: "In 1956, $X$ proved this and in 1957, $Y$ discovered that. If only I had been alert, I could have combined these results with ideas of my own and with Heisenberg's uncertainty principle and I could have done thus and such. Instead, $Z$ combined the ideas in a different way; for this work he later won the Nobel Prize."

I found this article to be an inspiration in several respects. First, I was amazed that anyone could understand his subject so well that he could recombine its parts in ways that never actually occurred. Second, I was amazed by Dyson's candor. Third, Dyson helped me see what creativity is. Fourth, he gave me a sense of the scope of knowledge.

Dyson's article is a sterling example of what good exposition can do. In a lifetime, you probably will not read more than half a dozen articles that are this good. But half a dozen is enough. If you are truly fortunate, and extremely talented, then perhaps you will write one.

# Chapter 4

# Other Types of Writing

*Stand firm in your refusal to remain conscious during algebra. In real life, I assure you, there is no such thing as algebra.*

Fran Lebowitz

*Neither can his Mind be thought to be in Tune, whose words do jarre; nor his reason in frame, whose sentence is preposterous.*

Ben Jonson
*Explorata—Timber,*
or Discoveries Made upon Men and Matters

*The flabby wine-skin of his brain*
*Yields to some pathological strain,*
*And voids from its unstored abysm*
*The driblet of an aphorism.*

The Mad Philosopher, 1697
in *The Devil's Dictionary*
by Ambrose Bierce

*What is written without effort is in general read without pleasure.*

Samuel Johnson

*Sometimes a cigar is only a cigar.*

Sigmund Freud

*The good writing of any age has always been the product of someone's neurosis, and we'd have a mighty dull literature if all the writers that came along were a bunch of happy chuckleheads.*

William Styron
interview, Writers at Work (1958)

*Close your eyes and think of England.*

—a Victorian mother, giving advice to her daughter
concerning behavior on the wedding night.

# 4.1   The Letter of Recommendation

## 4.1.1   Writing a Letter

Once you have become an established mathematician, you are likely to be asked for letters of recommendation. Such a document could be a letter of recommendation for a tenure case, or for a promotion, or for both. It could be a letter recommending a young person for a first or second job. It could be a letter recommending a senior person for an endowed Chair Professorship, or for the Chair of a department. (For the sake of this discussion, I will call these "professional letters.") It also could be a letter of recommendation for a student (such letters are treated a bit differently from professional letters—see below). There are many variants; here I would like to distill out some unifying principles on writing letters of recommendation.

When you are asked to write a letter as described in the first paragraph, you are in effect set a task. You have become a one-person "taskforce." What makes a taskforce different from a committee is that a taskforce is not supposed to debate the task at hand; instead, the taskforce is supposed to perform the designated task. In the present instance, you are supposed to offer in writing your professional opinion on a certain matter.

## 4.1.2   Responsibility of the Letter Writer

In my view, it is both unprofessional and irresponsible to dodge the assigned task. Let me be more precise. There certainly will arise circumstances where you either cannot write or should not write. Perhaps you have had a fight with the candidate in question and feel that you cannot offer an objective opinion; perhaps you have a conflict of interest; perhaps you are unfamiliar with the general area in which the candidate works; or perhaps you do not know the candidate well at all. In any of these cases, or similar ones, you should quickly and plainly write to the person (the Dean or Chair) who requested the letter and say that you cannot write it. Best is if you can give the reason, but it is acceptable if you cannot. Do not agonize over the task for six months and *then* decline to write; take care of the matter right away. The person soliciting the letter needs to collect the letters in a timely fashion; you are doing him/her no favor to artificially prolong the process.

The circumstances described in the last paragraph should be considered to be extreme exceptions. They will come up only rarely. In most instances, you will be asked to write about some particular person for some particular circumstance, and you should say "yes" and then you should do it.

I know mathematicians who will agree to write an important letter and then not do it. This paradox usually occurs for one of two reasons: (**1**) the putative letter writer is pathologically disorganized and forgets, or (**2**) the putative letter writer has nothing nice to say about the issue or person at hand and does not want to say it. I have already addressed the second of these *conundra*. The first of these situations is not likely to arise if the request to write was submitted to you as a formal letter—from a Dean, for instance. For then the piece of paper is sitting somewhere on your desk and you will probably get to it eventually. The paradox *can* occur if instead a student pokes his/her head in your door and asks for a recommendation to graduate school. You give a cheery "yes" and then the entire matter vacates your head. To avoid this error, ask the student to put the following on a slip of paper: his/her name, any classes he/she took from you or other pertinent data, and the address to which the letter is to be sent. (This ruse also helps you to avoid the embarrassment of having to ask the student's name.) Now you have it in writing. Also ask the student to come back in a week or so and remind you. I usually find it convenient to write the letter right away (if the student has poked his/her head in the door, then it is likely my office hour and I might as well be doing *something*). For once the request has been tendered, I am probably already thinking about what I am going to say; I might as well write it down and be done with it.

### 4.1.3   Getting the Job Done

Some clever people create a web page for students who want a letter of recommendation. Then, when a student comes to you and asks for a letter, you tell him/her to fill out the web page. This web page asks the student for all sorts of pertinent information—about courses taken, grades received, personal interactions, and so on. If you construct this web page carefully, then the student in effect writes the letter for you.

Make a point of writing requested letters in a timely fashion. It is the professional thing to do, and you would appreciate such consideration if the letter were about you.

Having decided to perform the task—to write the requested letter—you must do what you have been asked to do. You must formulate an opinion, state it clearly, and defend it. The standard format will be explained below.

### 4.1.4   What to Say in the Letter

In the first few sentences, state plainly the question that you are addressing. For example:

> The purpose of this letter is to support the tenure and promotion of John Q. Smith. I have known the candidate and his work for a period of six years, and have been impressed with his originality and his productivity. I indeed think that tenure and promotion are appropriate. My detailed remarks follow.

Alternatively:

> You have asked for my opinion on the tenure, and promotion to Associate Professor, of Dr. Smith B. Jones. Dr. Jones is now six years from the Ph.D., and in that time has produced nothing but some rotten teaching evaluations and a letter to the editor of the

> *Two-Year College Math Journal.* Based on that track record, my
> opinion is that he is worthy of neither tenure nor of promotion.

The bulk, or body, of the letter follows, and it should support in detail the
thesis enunciated in the first paragraph. I shall comment below on what might
constitute that support. First, let me conclude these beginning thoughts.

### 4.1.5   Conclusion of the Letter

Once the body of the letter is written—and this could comprise one or two
(or even more) pages—then you must write a concluding paragraph. You *must*
write it. You must sum up the point you have made, and restate your thesis.
A sample of this practice is

> In view of the stature of Jones S. Brown in the field of computa-
> tional algebraic geometry, and considering his accomplishments as a
> teacher and as a scholar, I can recommend him without reservation
> for promotion and tenure in your department.

(I am assuming that you have in fact described Brown's status and accomplishments, in a favorable manner, in the preceding paragraphs.) Another possibility
is:

> In sum, I feel strongly that Brown J. Smith should not be pro-
> moted or tenured. Indeed, I cannot imagine the circumstances in
> which such a move could be considered appropriate.

There are those who, although experienced letter writers, do not adhere
to the general scheme just described. One of the standard rationales for this
behavior is that, in many states and at many institutions, it is (theoretically)
possible for the candidate to have access to the complete text of his/her letters
of recommendation—including the identities of the writers. If such is the case,
then the soliciting school will inform the writer at the time the letter is solicited.
Of course the letter writer is offered the option up front of declining to write if
he/she is uncomfortable with this "freedom of information" situation.

### 4.1.6   Negative Feelings

There are those who, still uncomfortable, agree to write but are afraid to say
anything. The most negative thing that they are willing to do is to "damn with
faint praise." Not only does this artifice undercut the responsibility of the letter
writer, but it puts on those evaluating the case the onus of trying to figure out
what the writer was trying to (but did not) say. In the best of all possible
circumstances, someone at the soliciting institution will phone the letter writer
and just *ask* what the letter was meant to say. In the worst of circumstances,
the evaluators are left to guess what was meant. Given that someone's life and
career are in the balance here, it is a genuine shame for such a circumstance to
come to pass.

### 4.1.7 The Body of the Letter

Enough preaching. I will now give some advice about the body of the letter. If you want your (professional) letter to have some impact, and to be taken seriously, then you must do several things:

(i) Make some specific comments about specific work or specific papers of the candidate.

(ii) Make specific comments about the candidate's abilities.

(iii) Make specific comments about the candidate's standing in the field.

(iv) Make binary comparisons.

You may also wish to discuss other qualifications of the candidate. If you have first-hand knowledge of the candidate's teaching, then it may be quite appropriate to comment on that. If you have collaborated with the candidate, then you can say something about that. No matter what the topics may be, you should heed these principles: be *precise*, speak of *particular* attributes, and speak only of those topics of which you have *direct knowledge*. Now let me explain.

Your letter had better say more than "Smith J. Smith is a hail fellow, well met. Give him whatever he wants." First, such a letter does not say anything. Second, given the circumstances described above, in which some letter writers attempt to avoid litigation by "damning with faint praise," such a vague letter could be construed as *sotto voce* damnation. If your comments are instead detailed and specific, then it is difficult for people to misconstrue them.

Thus you should dwell, for a page or more, on specific virtues of the candidate's scientific work. Make detailed remarks about specific papers: Why is this result important? How does it improve on earlier work? How does the work advance the field? Who else has worked on this problem? This material should not be a self-serving introspection. Remember that most of the readers of the letter will be nonspecialists. Many, including the Dean and members of his/her committee, will not even be mathematicians. Thus attempt, briefly, to give background and motivation. Drop some names. For example, say that Ignatz of M.I.T. worked on this problem for years and obtained only feeble partial results. The candidate under review murdered the problem. If appropriate, point out that the candidate publishes in the *Annals* and *Inventiones*—and that these are eminent, carefully refereed journals.

### 4.1.8 Binary Comparisons

It is astonishing, but true, that even highly placed people, who write dozens of influential letters every year, seem to be unaware of the need for binary comparisons. To put it bluntly, an important letter which will have a strong effect *must* compare the candidate being discussed to other people, of a similar age and career level, at other institutions. The comparison should be with people—preferably other academic mathematicians—whose names the informed reader will recognize. Thus, if the candidate is an algebraic geometer and you say in your letter that "this candidate is comparable to David Mumford when Mumford was the same age," then most algebraic geometers will know exactly what you mean and will be extremely impressed; they will in turn explain to their colleagues the significance of your remarks. If instead you say "this candidate is comparable to Prince Charles when Charlie was a student at Gordonstoun," then nobody will know what you are talking about—and you can be sure that they will not be impressed.

To come to the point, if you are writing an important letter that you want people to notice, then you must say something like

> The five best people under the age of 35 in this area are $A$, $B$, $C$, $D$, $E$.

In the best of all circumstances, the candidate under consideration in your letter is one of $A$ through $E$—and you should point out that fact. Alternatively, you could say

> Two of the best people in this field, at the beginning tenure stage, are Jones and Schmones. Candidate Bones fits comfortably between them. Bones is surely more original than Schmones and more powerful than Jones.

Or you could say that the candidate falls into the next group. Or that the candidate is so good that it would be silly to compare him/her to the usual five best. Say what you think is appropriate. But *say something*. If you do not, then the readers will notice the omission and infer that, between the lines, you are saying that this guy is not any good. Better to say that he/she is number 15 than to say nothing at all.[1]

Tailor your binary comparisons to the circumstances. It would be inappropriate to compare a candidate two years from the Ph.D. with a sixty-year-old member of the National Academy of Sciences (unless the comparison is favorable, and you are trying to knock the reader's eyes out). It would probably be inappropriate to compare *anyone* with Gauss (although I *have* seen favorable comparisons with Gauss!). Note also that, if you are recommending a senior person for (just as an instance) an honorary degree, then binary comparisons

---

[1] A *caveat* is in order if the letter that you are asked to write is *not* solicited from a research institution. If the candidate is in fact at a four-year college, where the primary faculty activity is teaching, then the school probably demands a lot of classroom activity—and not so much scholarship. These days, almost every school wants its permanent faculty to have some sort of academic profile; but a teaching college can hardly expect its instructors to stand up to hard-nosed binary comparisons. The lesson is this: read the soliciting letter carefully; speak to people in *their language*, and tell them what *they* want to know. If the soliciting letter is from a teaching institution, then it is probably most appropriate for you to write about teaching, curriculum, publications in the *Monthly*, and letters to the editor of *UME Trends*. A disquisition on Gelfand-Fuks cohomology is probably less apropos.

might be entirely out of place, and uncomfortable as well. If the person is already a Chair Professor at Harvard, then to whom will you compare him/her? And to what end?

### 4.1.9 Other Specifics in the Letter

Your letter of recommendation can contain other specifics and details that might grab the reader's attention. You could say that the candidate gives excellent talks at conferences. You could say that he/she is a wonderful collaborator. You could say that the candidate has beautiful insights, and that talking mathematics with this person is a pleasure.[2] You could describe in glowing and heartfelt terms the process of proving a theorem, or of writing a paper, with the candidate.

These days, credible evidence that the candidate is a good teacher will certainly help the case. Of course you are probably not in the same department as the candidate, so you very well may not be able to discuss his/her teaching. If the candidate is a truly outstanding teacher, then perhaps you have heard his/her colleagues mention his/her talents, or perhaps you know that he/she has won a teaching award. Maybe you have heard the candidate give an inspiring talk at a conference. It makes quite an impression on letter readers if Professor $A$, from University $X$, can comment knowledgeably and in detail on the teaching of Professor $B$ from College $Y$.

### 4.1.10 Travesties in Letters

Here are some travesties that I have seen (all too frequently) in letters of recommendation. You should certainly not emulate any of these mistakes:

1. The writer begins in one of the fashions indicated above. Then he/she says

   > Mary P. Smith has proved the following theorem about pseudo-graphs (state the theorem). This is a nice result. The theorem is based on some old ideas of mine. [*And the rest of the letter consists of a description of the letter writer's own work!*]

   Such a letter violates all the precepts laid out above, and marks the writer as a thoroughly self-absorbed fool. Of course this letter does nothing to help, nor to hurt, the candidate; but it gives a rather poor impression.

2. The writer discusses the candidate, discusses the candidate's work, makes binary comparisons, and mentions specific papers. In short, the writer makes all the right moves. In the concluding paragraph, he/she says

---

[2]I saw one letter of recommendation, by a very famous mathematician about another famous mathematician, that said, "Talking mathematics with $X$ is like talking to Enriques." This written by someone who was too young to have ever interacted meaningfully with Enriques.

I am going to make no specific recommendation as to whether you should promote Smith J. Brown or not. After all, you know what the needs of your department and your school are. You can use the information that I have provided to come to an appropriate decision.

Rubbish! Imagine taking your car to a mechanic and hearing him say "Your transmission runs at half speed and your rear wheels turn forward. Your stroke is short and your valves rattle. I am not going to make any specific recommendation for repairs because, after all, it is your car and you know what your needs are." Or imagine your physician saying "Your heart will give out any day now, and you are also a prime candidate for a stroke or for total paralysis. However, I will make no specific recommendations. It is your body, and you know best . . . ." *You are a professional; you are expected to render an opinion.*[3]

3. The writer neglects to address explicitly the question at hand. This omission is sometimes committed inadvertently, but this omission is a dreadful error. If you are asked whether Jonesy B. Smithy should be tenured, or promoted, or given a certain post, or a grant, then you must say point blank what your opinion is *about that question as it applies to that candidate.* If you neglect to say, then your letter (taken as a whole) is likely to be read as the worst sort of "damning with faint praise." Whether you intended it or not, you may have buried the candidate.

4. The writer faces the following request (and blows it): In a school that fancies that it wants to make hard decisions, and elicit the *bona fide* truth from the letter writers, it is common for the Dean to include in his/her solicitation letter a query like "Would you tenure Barbara Jones in *your* department?" If the person being asked for the letter works at Harvard, and if the institution soliciting the letter is a four-year teaching college, then such a Dean is just looking for trouble.

Even if the letter of solicitation does not explicitly ask this question, we letter writers are often tempted to answer it. Unfortunately, the answer sometimes comes out like this:

(∗)    Dr. Brown P. Smith is not good enough for us,

---

[3]I must confess that not everyone agrees with me here. I have spoken to Deans who noted that an outside letter writer may know a lot about the research of the candidate, but probably knows little or nothing about the teaching or departmental service of the candidate. The letter writer also probably knows precious little about the value system of the school in question. So how can he/she make a recommendation for or against tenure (with such incomplete information)? All I can say is that, when I was Chair and had to present tenure cases to the Dean, the Dean would certainly take note of letters that failed to make an explicit recommendation. And he did not like such letters. If you want to be extra careful, then you can make it clear in your letter that you are basing your recommendation for or against tenure *on the explicit information that you have.*

but he is certainly good enough for you.                    ✠

Rarely is a letter writer clumsy enough to phrase things quite this bluntly, but I have seen many a letter in which this sentiment comes through loud and clear.

This is just too bad. The person writing such a statement (or a euphemistic paraphrase of it) probably thinks that he/she is being frank and helpful. He/she is being neither. Instead, he/she is insulting the maximum number of people in the least constructive possible fashion. A word to the wise should be sufficient: proofread your letter of recommendation to be sure that you have not inadvertently (or intentionally) made statement (∗). The inclusion of such an assertion in your letter will vex the readers, and render your letter ineffectual, so that it will not count. I presume that this effect is not the one that you want.

If in fact you are at a place like Harvard, and if your letter is solicited from a much more humble institution, and if you *must* address this difficult question, then you should endeavor to tell the truth. Say that Harvard's math department is usually ranked in the top three; you only tenure people who are world leaders, indeed great historical figures; such standards would be inappropriate to apply at an institution like the one which has solicited the present letter. However, you certainly would recommend this candidate for tenure at Bryn Mawr, or Swarthmore, or some other institution that you choose for comparison.

## 4.1.11 Responsibility of the Letter Writer

That concludes my enumeration of woeful errors. Now let me cut to the chase. When you are writing a letter for a candidate, then a heavy responsibility rests on your shoulders. The Dean or Chair who solicits the letters of recommendation is not simply casting his/her net and taking a vote: this person wants a *mandate*. He/she will *not* weigh good letters against bad: he/she wants to be socked between the eyes. A tough Dean once told me "If a case is not overwhelming then I turn it down. If the candidate is any good, then he/she will land on his/her feet. If not, then we are better off without him/her." Thus if your letter says

Smith P. Smith is no good. Don't do it.

then you may as well face the music and realize that *your letter alone* will have killed the case—at least for now. I cannot repeat this point too strongly: it is dead wrong to say to yourself "This is a negative letter that I am writing, but it will not count unless all the other letters are negative too." Baloney! One negative letter will usually stop the case cold. That is all there is to it.

A letter with inadvertent errors (of the sort mentioned above) will not necessarily bury the candidate, but it certainly will not help him/her.

## 4.1.12  The Closing Paragraph

In the closing paragraph of your letter, you will typically indicate a degree of enthusiasm for the case at hand. Here is a graded list of examples—taken from letters that I have actually seen:

> Jones P. Jones has done a workmanlike job with his research program.

> Smith P. Smith is a reasonable case for tenure. You would not go wrong to tenure him.

> I recommend Brown P. Brown warmly for tenure and promotion.

> I recommend Jones P. Brown enthusiastically for tenure and promotion.

> The case for Brown P. Jones is overwhelming. I recommend him without reservation.

> I give Smith P. Jones my strongest possible recommendation. Phone me if you require further details on the case.

In case my admonitions have not sunk in, let me beat you over the head with them. The first two of these statements are in fact negative. Whether they were *intended* to be negative, or are simply an articulation of the writer's loss for words, this is how they will be read. You might as well take the candidate out and shoot him. The third passage is a little better (many evaluators will read "warmly" as "lukewarmly"), but does not convey passionate affirmation.

By contrast, the fourth example will definitely be construed favorably. The adverb "enthusiastically" conveys the positive nature of the assertion. The last two samples represent the sort of forcefulness considered to be virtually mandatory if you want to argue for the tenuring of a candidate at any of the best institutions.

## 4.1.13  Letters Are Not Formulaic

The writing of letters of recommendation is not formulaic. Indeed, if all letters of recommendation fit a pattern and sounded the same, or if all *your* letters look the same, then they will eventually be ignored. Mathematicians keep a mental database on letter writers in the same way that good baseball pitchers keep a database on batters. After several years, we know who "tells it like it is" in his/her letters, who spins tales, and who simply cannot be trusted. We know who always writes the same letter for everyone. And we act accordingly.[4]

You will develop your own style of writing letters. Mathematics is a sufficiently small world that, after several years, people will recognize your letters of recommendation at a glance. But, no matter how you write your letters, you will want to take into account the issues raised in this section.

### 4.1.14 Enthusiasm in Letters

During times when jobs are hard to come by, letters of recommendation tend to become more and more inflated. Everyone feels that he/she must try harder if he/she is going to land a job for that special someone. Here are examples of lines that have actually been used to describe specific, rather famous, job candidates. I do not necessarily recommend that you use any of them; if you do, the readers might think that you are eating with only one chopstick. But these examples will give you an idea of what some people have done to draw attention to what they are saying, or to remove their particular letter from the ranks of the humdrum. (Of course names have been changed to protect the innocent.)

> Smith P. Smith has a good idea every other day and writes a brilliant paper every week.

> Brown P. Brown knows both classical analysis and modern analysis. He is the natural successor to Hardy and Littlewood.

> Talking to Jones P. Jones is like talking to Enriques. (An inspiring thought, written by one too young to have ever interacted meaningfully with Enriques.)

> Jones P. Brown is the most mathematically intelligent person that I have ever met.

> Brown P. Jones is the greatest mathematician since Gauss.

### 4.1.15 Letters for Students

Although there is an art to writing a "professional letter," it is also the case that at least you are dealing with familiar territory, and speaking of matters on which you are expert. Any professional mathematician for whom you might

---

[4]In fact there is an eminent mathematician who has had many students and writes a great many letters of recommendation. They are so similar that you could hold any two of them up to the light, one behind the other, and most of the words would line up. But then he scribbles his real opinion in the margin by hand.

write has a publication list, and a track record in teaching, and a reputation as a lecturer, and some *gestalt* as a collaborator. When you are writing for a student, by contrast, matters are more nebulous. The student has none of the professional attributes that you are comfortable discussing. Yet, if you want your letter to be memorable, and to be perceived positively, then you still want to say something noteworthy about the student.

While the precepts of organization that I have stated above still apply in a letter for a student, some of the other particulars do not. For example, you most likely cannot remark on the student's scientific work, and you most likely cannot make binary comparisons. In fact any attempt at binary comparison is likely to be absurd. Imagine saying "I am delighted to recommend Smith T. Smith. She is every bit as good as Brown Q. Brown, whom I recommended five years ago to a different institution." If in fact you previously recommended a student who turned out to be a well-known star—or at least a well-known star at the institution to which your letter is addressed—then by all means make a binary comparison involving that person if such a comparison is appropriate. Usually it is not appropriate, so no such comparison should be included.

Thus in practice you must try a bit harder to say something specific about the student for whom you are writing a letter. After you have been teaching for several years, it may be the case that you have actually taught a few thousand students (this would be true, for example, if you have taught calculus in large lectures several times). It becomes difficult to distinguish students—even good ones—in your memory, much less to say something of interest about any of them.[5] If you apply yourself to the task, then you can nevertheless come up with some noticeable things to say. Here are some examples, taken from genuine letters:

> Roman P. Jones is one of the five most talented undergraduates that I have encountered in twenty years of teaching.

> Jones P. Roman is hard working and perseverant. She can think on her feet—at the blackboard—just like a mathematician. She is original and imaginative.

> In order to test her creative abilities, I have given Meredith Smith extra work outside of class. She discovered a new proof of Gronwall's inequality, discovered Euler's equation in the calculus of variations on her own, and has also posed numerous interesting problems of her own creation. Needless to say, she breezed through all the standard class work.

---

[5]I was an undergraduate at the University of California at Santa Cruz (UCSC). At UCSC we had no grades; instead the professor was supposed to write a paragraph about you and give you either a "Pass" or a "Fail." It was generally hard for the professors to think of anything to say in these paragraphs, especially because the professors often got several terms behind in doing their writeups. The result was a lot of misinformation and disinformation.

As usual, the point is to say *something*—and that something should be quite specific. The view of letter *readers* is that if the letter writers cannot say anything unambiguous and remarkable about a student, then there is probably nothing remarkable about that student. So what if the student can earn mostly As in his/her classes?—this fact is no big deal, and in any event can be gleaned from the transcript.

### 4.1.16 Declining to Write

Sometimes a student, or someone else, will ask you for a letter about himself/herself and you do not feel that you can write a good one. Either you have nothing to say, or you have nothing good to say, or you have some other valid reason for not writing. (Note that this case is different from the one in which a Dean is asking you for a letter about one of his/her faculty. Now the candidate himself/herself is standing before you and asking for a letter *about himself/herself.*) You always have the option of agreeing to write, and then writing a negative letter. Often, however, you bear the candidate no malice and think that he/she deserves a chance. In that case, the honorable thing to do is to say to the candidate "I'm sorry. I frankly do not feel that I could write a good (or supportive) letter for you. Perhaps you should ask someone else." The rotten thing to do—and this happens far too often—is to say "Oh yes, fine" but with no intention of ever writing *anything*. Note that the lack of your letter in the dossier will make that dossier incomplete; in many cases the candidate will not, as a result, be considered at all. If such is the effect you want, then you should have the courage to say something in a letter. If it is not the effect you want, then you should have the courtesy to take a "pass."

### 4.1.17 Letters for Your Masters and Doctoral Students

One of the most critical, and delicate, types of letter that you will have to write is a letter seeking a job for a student completing his/her M.S. or Ph.D. under your direction. Your statements are *a priori* suspect because you obviously have a vested interest in finding this student a job, and in seeing him/her succeed. Thus you must strive to put into practice the precepts described above: **(i)** say why this student is good, **(ii)** say what this student has accomplished, **(iii)** if possible, compare the student favorably with other recent degree holders, **(iv)** say something about the student's ability to teach, **(v)** say, if appropriate, something about the student's facility with English, **(vi)** say something about the student's ability to interact with other mathematicians.

A meat-and-potatoes job application from a fresh Ph.D. has a detailed letter from the thesis advisor that conforms, at least in spirit, to the suggestions just adumbrated. This detailed letter is accompanied by two or three additional letters from other instructors at the same institution, each of which is rather vague and says in effect "Doo dah, doo dah; see the letter by the thesis advisor." If you want your student's dossier to stand out, and to really garner attention, then you should strive to help the student make his/her dossier rise above this rather dreary norm. Endeavor to ensure that the other writers know something about what is in the thesis. If possible, convince someone from another institution to write a letter for the student. Make sure that the dossier includes detailed letters about the student's teaching abilities.

### 4.1.18 Truth in Letters

When you write a letter of recommendation, tell the truth. If all your letters read "This candidate is peachy, and a dandy teacher too. Give him/her $X$" (where $X$ is the plum that the candidate is applying for), then after a while nobody will pay any attention to what you say. I presume that if you take the trouble to write letters, then this is not the result that you wish. The infrastructure has a memory. It will remember whether you are a person who can make tough decisions, or whether you are wishy-washy. If you want your letters to count, then you must call it as you see it. It is hard to be hard, but sometimes the situation demands it.

I am rather good at writing letters, and I have a certain reputation in this regard. As a result, I am asked to write quite a few letters, and they really seem to make a difference. I have helped a number of people to get good jobs, I have helped a number of people to get tenure or get promoted to full Professor or be awarded endowed Chair Professorships. I have also made it happen that certain people *did not* get tenure or *did not* get promoted. I feel as though I have had some impact on the world, and it is a satisfying feeling.

### 4.1.19 Ability with English

One issue that we, as letter writers, often must address is whether or not a job candidate can speak English, and how well (this question could even apply to an undergraduate student—especially if that student is applying to graduate school and might be considered for a Teaching Assistantship). In this matter we are, in the United States, cursed by our group dishonesty over the past forty years. Too often have we said in a letter that "this candidate speaks excellent English, can teach well, and is a charming conversationalist to boot." In a more frank mode, we might have said "This candidate speaks better than average English" (recalling Garrison Keillor's statement about the town of Lake Woebegon, in which "all the children are above average"). When the candidate arrived to assume his/her position, the hiring institution often found that he/she could not understand even simple instructions and had no idea how to teach.

It is difficult, but you must endeavor to be honest about the candidate's fluency in English (again, your credibility—which will follow you around all your life—is at stake). You could say, for example,

> This candidate speaks English like a home-grown American, with no trace of an accent. Listening to him/her is like listening to Donald Trump.

This would be the ideal thing to write, and would dispel all trepidations about the candidate's fluency. Unfortunately, if the question needs to be addressed at all, then this statement probably is not true. You could instead say

> Gronwall Bronsky has been taking "English as a second language" and has taught several lower-division courses successfully. His English is accurately formulated and clearly enunciated. Students have no trouble understanding him.

Unfortunately, you cannot always be so enthusiastic. Sometimes you must say something like

> Ms. Ching P. Chang has been working hard on her English, and has made substantial progress. One still needs to concentrate in order to understand her.

Or you might say

> It takes students three or four days to become accustomed to Ms. Imelda Marcos's English, but her charming personality helps them along. As a result she is a most successful teacher.

The thought that I am trying to formulate here is that Ms. Chang's English or Ms. Marcos's English is not perfect. But Ms. C and Ms. M are real troupers. They try hard, and the students (at least in Ms. M's case) forgive them a lot.

Of course you can plainly see that I am trying to suggest ways to avoid saying "This person cannot speak English and refuses to learn. He/she is only suitable for a nonteaching position." But sometimes—presumably not in the case of Ms. Chang or Ms. Marcos—it must be said.

At the risk of repeating myself, let me say that when you address the candidate's ability with English, then you should not be formulaic. If all your letters about foreign candidates say

> X's English is just fine. He/she is a good teacher.

then, after a while, the world will nod out by that portion of your letters. Try to say something original, apt, and true about each candidate. I once wrote the following about a fresh Ph.D., from a foreign country, who was applying for a job:

> I consider myself to be rather a good teacher, but I really learned something when watching Mr. Marcel Duchamp with his class. He moves skillfully among his students, looks at their work, makes insightful remarks, and does a marvelous job of eliciting class participation. It is clear that the students like and respect him.

What I said here was absolutely the complete and honest truth about my opinion of this candidate's teaching (and implicitly his English) ability. I cannot help but think that this made a very positive impression. The passage addresses the language issue implicitly, for it confirms that the candidate can *teach*. Moreover, it is not just a bunch of pap. It says something particular and notable about the candidate's abilities.

## 4.1.20 Thorny Matters

Occasionally, you will have to address a truly thorny matter in one of your letters of recommendation. As an instance, I was once writing on behalf of a young mathematician who was applying to several dozen first-class universities for a position. I thought that I knew this person quite well. But, a few days before I was going to draft my letter, I learned that the candidate was undergoing a sex change. I had to decide whether I should mention this fact in my letter. I reasoned as follows: if he were changing from Catholicism to Judaism, or from Democrat to Republican, or from carnivore to vegetarian, I certainly would not consider discussing the matter in my letter of recommendation for a mathematical post; so why should I treat transsexuality? And I did not. Some time later, I discussed the matter with one of my mentors. He told me that I had erred. In stern terms, he informed me that a matter like this could affect the candidate's ability to teach, and his ability to function as a colleague; therefore I was morally obligated to mention the matter. I still do not know what the correct course of action should have been. I only hope that I will not be faced with another choice like this one any time soon.

## 4.1.21 Miscellaneous Issues

And now a few miscellaneous issues. Many talented young mathematicians do much of their early work with a collaborator. Often the collaborator is the thesis advisor. It is somewhat natural for someone evaluating the merits of such a person to wonder what percentage of the work under review is due to the young person and what percentage is due to the senior co-author. The problem is exacerbated if the young person's surname is something like Zymurgy. Then, because the custom in mathematics is to list authors alphabetically by surname, this budding young mathematician will always be listed last.[6] It is probably appropriate in such a situation for you as letter writer to explain how the world works, and why Zymurgy's name always comes last. Another unfortunate possibility is that the institution soliciting the letter of recommendation will ask you point blank how much of the work was done by the young candidate. I find such requests irritating, and I tend to avoid giving a direct and/or analytical answer. I will often say something like, "I have spoken with both authors about this work and it is clear to me that they both contributed significant ideas."

Just for fun, let me conclude this long section by quoting from a letter for tenure that was written (truly!) about thirty-five years ago for a candidate in a French department. Call the candidate Mr. de Gaulle.

> Surely Mr. de Gaulle is now wiser than he once was.

*That was the entire text of the letter!—No introduction, no conclusion, no binary comparison, no exegesis of the candidate's scholarly work. Just the one sentence.*

---

[6]We must note here that, in many other fields, like medicine, the custom is *not* to list authors alphabetically by surname. Instead authors are ranked according to how much they contributed to the work. Also the person who runs (and funds) the lab gets a top billing.

Although the letter does not follow the precepts described in this section, it definitely gets its point across.

## 4.2 The Book Review

### 4.2.1 How to Review a Book

As with most topics in the subject of writing, there is some disagreement over what constitutes a good book review. When Paul Halmos was the book reviews editor of the *Bulletin of the AMS*, he sent every reviewer a set of instructions. The gist of these instructions was that a book review is not a book report. It should *not* say "Chapter 1 says this, while Chapter 2 says that. Chapter 3 is a bore, and Chapter 4 is too hard."

Instead, according to Halmos, a book review on a book about $X$ is an excuse to write an essay about the subject $X$. Look at the book reviews in the *New York Review of Books*. On the whole they are a delight to read, and they conform to Halmos's view of what a book review is and does. These reviews tell you about the book, but they paint the picture on a large canvas.

To reiterate: If you are reviewing a book on harmonic analysis, then you should write about the history of the subject, what the milestone books and theorems have been, who the major players are and were, and what the big problems are. Drop some names. Make some assertions and conjectures. Having laid considerable groundwork, then finally focus on the book under review. Describe where it fits into the infrastructure you have outlined. Indicate its strengths and weaknesses. Suggest who would profit from reading it, and why. Touch on areas where there is room for improvement. Do not, however, use my suggestions here as an excuse to write an opinionated essay and virtually ignore the book. The book review is supposed to be *about the book;* but it should be about the book in the context of the subject matter, not the other way around.

### 4.2.2 Desiderata for a Book Review

Here are some other issues that your book review might address:

- Will students benefit from reading the book?
- Are there exercises?
- Are there lists of open problems?
- Is there an accurate and complete bibliography?
- Is there an index?
- Is there a list of notation?
- Is there a glossary?
- Are there good and useful figures?
- Are there an adequate number of examples and are they stepladdered?

- Is there sufficient review material? Does the book begin at a reasonable level?

- Does the author provide an adequate amount of detail in the book? Does the book make too many demands on the reader?

- Are the proofs complete, clear, and accurate?

- Is the book organized in an intelligent fashion that is useful to the reader? Can the beginner navigate his/her way through the book?

- Is the history correct? Are attributions complete and accurate?

- Does the book bring the reader up to the cutting edge of research?

If you think about the issues that I have raised here, then you will realize that I have described what a potential reader of the book will want to know when he/she is making a decision as to whether to buy the book and whether to read the book. Or perhaps he/she wants to know whether to recommend the book to students or colleagues. One of the main purposes of your review is to inform such decisions.

### 4.2.3   Positive and Negative Reviews

Many mathematical book reviewers—writers for the *Bulletin of the AMS*, for instance—feel obligated to write a *positive* or upbeat book review, no matter what they really think of the book. They are afraid to be critical. In my opinion, this attitude is an error. Not all books are good, and not all good books are entirely good. You will help the audience, and the author as well, if your review points out inadequate features of the book, or omissions, or errors, or items that can be improved. You should tender your criticisms in a constructive and civilized fashion: in this manner you will increase the likelihood that people will attend to what you have to say, and your thoughts will perhaps make friends and influence people (rather than the opposite).

On the other hand, there is the occasional reviewer who lets it all hang out. Books seem to have a sort of permanence that papers do not. An incorrect or wrong-headed paper is, after all, ultimately buried in a bound journal volume and hidden away. But a book is always right there on the shelf, staring us all in the face. And, as previously noted, a book reaches out to a larger audience than does a paper. As a result, emotions can run high over a book. I have seen a book review that (literally) began by questioning the editorial decision to publish the book and asserting further that the book completely misrepresented the subject matter; the reviewer spent the rest of the review describing what the subject was *really* (in his opinion) about, with nothing further said about the book itself. I have also seen a book review [Blo] that declared the math so beautiful, they compared the subject matter (a bit indelicately) to a woman, "In short, this is a gal all the boys should be pursuing...Enter the good doctor Milne with an impressive tome revealing all, including diagrams of her private parts." A recent (and rather controversial) book review [Kli] asserts that the book under review is obviously about a weak subject, as one can see by examining the Bibliography and noting the substandard journals in which the cited papers are published; the reviewer neglects to point out that he is or has been on the editorial boards of most of the relevant journals (see [NoS] for an incisive reply). While these essays are briefly diverting they are, in retrospect, embarrassing for us all. As you write your review, pretend that you are reviewing the book of a friend: you want to be honest, and you want to be helpful, but you also want to be scholarly and dignified. Brutality is almost never the order of the day.

In 1978 there appeared a marvelous book on algebraic geometry that is almost universally admired, but which is famous for having a large number of errors: either slight misstatements, or omissions, or incomplete proofs. The fact remains that everyone loves this book, and there is no other like it. (Heck, I may as well tell you: it is [GH].) One reviewer [Lip] praised the book to the heavens, but felt that he had to say something about the hasty writing and the density of errors. So he wrote in part

> If it makes you feel better, think of this book as a set of lecture notes, or even as a fantastic collection of exercises, with copious hints.

Thus the reviewer did his duty: he certainly said something critical, but he said it in a charitable manner, and with good humor. Even the authors must have chuckled over these remarks, and everyone learned something.

### 4.2.4  Harsh Book Reviews

The harshest book review that I have ever seen appears in [Mor]. Mordell in fact uses this review to trot out his frustration with the French school's abstraction of his beloved number theory. He attacks not just the book, but he attacks its author on a rather personal and visceral level. A now famous letter was subsequently written by C. L. Siegel to Mordell, praising the review and heaping even more calumny upon the book's author. A discussion of that interchange, and its significance, appears in [Lan]. The trouble with such a review is that any flow of scholarly thought or criticism is lost in the morass of venom and vituperation. No constructive purpose is served by such a review. It is also virtually impossible to have any useful dialogue following upon such a review.

If you are called upon to review a book, and you are tempted to trash it, then I suggest that you set the draft of your review aside for a month (a year would be too long!). Let the ideas gel, and let the words mellow. Show it to a few trusted friends. After a month, you will probably be inclined to take the long view, and to express your ideas in a more temperate fashion. The result will be a better review, and one that you will still be proud of ten years after it appears.

## 4.3   The Referee's Report

### 4.3.1   The Art of Refereeing

When you are asked to write a referee's report on a research paper, then you are being requested to offer your opinion as an expert. If you agree to write the report (and you *should*—refereeing is an important part of your professional duties), then you should adhere to the following precepts:

- Write the report in a timely manner—if possible within the time frame suggested by the editor.

- Tell it as you see it. Just as in a letter of recommendation, enunciate your opinion clearly and succinctly, defend it, and then summarize your findings.

- Defend your opinion in detail. You need not find a new proof of each lemma, nor read every bibliographic reference. But you must read enough of the paper so that you can comment on it knowledgeably. While you may not have checked every detail in the paper, you should at least be confident of your opinion as to the paper's correctness and importance.

  If somebody asked you whether you liked your car, and whether you would recommend that they buy one, you would not (in all likelihood) tell how each bolt was installed in the chassis, nor how the finish was applied to the body. You would instead summarize the overall performance and features of the automobile. Just so, when you evaluate a paper you should address Littlewood's three precepts: **(1)** Is it new? **(2)** Is it correct? **(3)** Is it surprising? You should speak to its contribution to our knowledge, and to the literature.

- Provide constructive criticisms of the writing, or of the paper's organization. You may enumerate spelling and grammatical errors (if you wish to do so). You should certainly point out mathematical errors, or places where the reasoning is unclear. But you should not be captious. (*Exercise:* Look up this word on `Google` and think about its relevance to the present discussion.)

- Place the paper in context: How does it compare to other recent papers in the field? Where does it fit? Does it represent progress? If you were not the referee, then is it a paper that you would want to read?

### 4.3.2   Tailoring Your Report to the Journal

Of course your report should be tailored to the journal to which the paper was submitted. For instance, the *Annals of Mathematics* professes to publish papers of great moment, written for the ages. Other journals have the more modest goal of publishing papers that are correct and of some current interest. Still other journals have no standards at all. You must speak to people in their own language—language that they will understand. When you evaluate a paper for a journal, base your assessment on *that journal's value system*.

A typical referee's report is one or two or, occasionally, more pages. Its most important attribute should be that it makes a specific recommendation. Everything else that you say is for the record: it is important, but it is secondary.

### 4.3.3 Reviewing a Book

The principles that I have described here for evaluating a research paper also apply when you are reviewing a book project. But there are other factors that come into play:

- Any scholarly book should have a thoughtful and well-written Preface. That Preface should tell the reader what this book is about, who is the intended audience, what are the prerequisites for the reader, and into what context the book fits.

- Any scholarly book should have a detailed and complete Index.

- Many scholarly books will benefit from a Glossary.

- Many scholarly books will benefit from a Table of Notation.

- Many (but not all) scholarly books should have exercises.

- Many scholarly books feature an essay at the end of each chapter explaining the context and provenance of the ideas in that chapter. This is called the "Princeton style."

- Most scholarly books should have well-rendered and carefully chosen figures.

Your review of a book project will typically be from 2 to 5 pages. The author of this book has clearly put a great deal of time and effort into this work. You should respect that fact. Even if you really do not like the book, formulate your remarks and criticisms in a constructive fashion that will be useful to the author and to the editor.

## 4.4 The Talk

### 4.4.1 How to Hold Forth

Giving a talk is different from writing. But it is relevant to the writing process. We ordinarily do some writing to prepare for a talk. Frequently the talk is about a paper that one has recently written. And what we write will strongly influence the talk itself. So this topic is fair game for the present book.

A talk is more flexible than a paper. In a talk, you may indulge in informalities, whimsicalities, and a little imprecision; it helps the audience a lot to tell of things tried, and things that failed. You may work trivial examples, and use them as a foundation on which to build ideas.

A talk is also less flexible than a paper. Because the audience receives the talk in linear order—it cannot rewind or speed ahead to check on things—it is therefore at your mercy. You are at a great advantage, when preparing a talk, if you are aware of the limitations of the medium. Endeavor to be gentle.

### 4.4.2   What Not to Do

John Wermer [Wer] makes an excellent case that many mathematics talks are not as effective as they might be because the lecturer is speaking to an imaginary audience located inside his/her head. This audience is one that knows all the necessary motivation, can pick up on fifty new technical definitions quickly and easily, can follow a technical proof (without explanation) in a jiffy, and can fill in the logical gaps and potholes left by the speaker. Of course such an audience is apocryphal, and thus we are often left with a communication gap between speaker and audience. This section will give you some advice on how not to be like Wermer's *idiot-savant*.

You want to be sensitive to your audience when giving a talk. Realize that these people, smart though they may be, can only absorb a couple of ideas in an hour. They cannot apprehend very many new definitions in a short time window. And also realize that you have a very limited amount of time in which to present your ideas. You must focus on the main point and get it across cogently, concisely, and imperatively. Do *not* give a lot of technical detail. Do not have an excess of fancy graphics. Or fancy anything else. Your talk should really be an advertisement for your ideas, not a very detailed presentation of your ideas. One way to think of your talk is that it is an invitation for those who are interested to buttonhole you afterward and have a serious conversation on the subject.

### 4.4.3   Ingredients of a Good Talk

What are the ingredients of a good mathematics talk? First, you must know your subject cold. This does not mean simply that you know it well enough to communicate it to another expert like yourself, but rather that you know it well enough to teach it: that you know the background, the biases, the reasons for the questions, the good and the bad attacks on the problems, and the current state of the art. However, just because you know all these things does not mean that you need to say all of them. A good mathematics lecture is an exercise in self-restraint. Never mind impressing the audience with your profound erudition, your spectacular vocabulary, your extensive professional connections, or your readiness to cite last week's hot results. Instead showcase a nugget of knowledge and insight, and shore it up with crisp comments and incisive examples.

### 4.4.4   Time Management

If your colloquium talk is scheduled for fifty minutes, then the first twenty should be accessible to a graduate student who has passed the quals. My statement is a strong one. Such a student is not expert at anything. He/she knows the basics of real and complex analysis, algebra, and perhaps a little geometry. This student has (we hope) an open mind and wants to learn. But your talk in those first twenty minutes should presuppose no specialized knowledge beyond what has just been mentioned. This explains why a nice example or two can be so useful. With an example, God is in the details. The playing field is level, and everyone can benefit. The example(s), of course, should lead to some definitions and the formulation of the questions that you wish to treat in the body of the talk.

The next twenty minutes of the talk should be pitched at a mathematically literate person who is not a specialist. By this I mean that, if your talk is about some part of analysis, then the second twenty minutes should be comprehensible *not just to a specialist in another part of analysis*, but to an algebraist. So you can mention more sophisticated ideas—sheaf theory, or elliptic regularity, or wave front sets, or singular integral operators—but you should not beat them to death.

The last ten minutes can be for the experts, for God, and for you (not necessarily in that order). Every speaker should have a chance to strut his/her stuff, and the end of the talk is when you should do so. Mention some gory details. Make speculations, formulate technical corollaries, sketch the key ideas in the proof. Forget the neophytes and address yourself to the people who might read your papers. In fact if you do not use the last ten minutes of your talk in this way, then you might leave the impression that you are a lightweight, or that you have nothing new to say.

One witty mathematician pointed out that the *last five minutes of a talk* (after the ten minutes noted in the previous paragraph) can be for the speaker and God.

### 4.4.5   How to Finish a Talk

Attempt to finish with a bang. Too many math talks begin with "Well, what I want to talk about today is . . . " and then a definition goes onto the blackboard. Too many math talks end with "Well, I guess that's all I have to say" or "I see that I'm out of time so that's it" or "I guess I'll stop here; thank you." Surely you can devise a more creative and informative way to conclude your discourse. You would never end a paper in this fashion. Of course when you write a paper you have time to sit and think of a nice turn of phrase for your conclusion. You should do the same when composing a talk: prepare the introductory sentence or two in advance; likewise prepare the concluding sentence or two. You could finish with a few courteous words of thanks for the opportunity to visit your hosts and to enjoy the hospitality of their department; or you could end with a few mathematical sentences—of real substance—that summarize your enthusiasm for your subject matter. But do end by *saying something*.

The preceding discussion may make it seem that giving a good colloquium talk is a piece of cake; that it requires only an acclimatization to certain simple proprieties. Not so. Many other parameters figure into the process.

## 4.4.6   Types of Talks

In fact there are many types of talks: the colloquium, the seminar, and the "job" talk (in which you are showcasing yourself before a department that is considering offering you a job) are three of these. The colloquium is supposed to be for the entire department and perhaps for the graduate students as well; the seminar is for a group of specialists, probably your friends; and the job talk is a set piece—something like Kabuki theater—in which you show yourself. An entire separate book could be written on the art of giving talks. In the interest of brevity, my remarks below will center around colloquium talks. Seminar talks are less demanding and job talks more so. The remarks below apply in some form to *any* talk; the trick in interpreting my advice for a particular circumstance is to *know your audience*. As you read my detailed remarks below, keep this unifying principle in mind.

## 4.4.7   Talk Checklist

1. *Showcase one theorem*, or perhaps a single cluster of theorems. There is no point to giving a talk on five truly different theorems, because the audience cannot absorb so much material in one sitting. On the other hand, if you cannot build your talk around one theorem, then perhaps you have nothing to say. Here is what Gian-Carlo Rota thought about the matter:

   > Every lecture should state one main point and repeat it over and over, like a theme with variations. An audience is like a herd of cows, moving slowly in the direction they are being driven towards. If we make one point, we have a good chance that the audience will take the right direction; if we make several points, then the cows will scatter all over the field. The audience will lose interest and everyone will go back to the thoughts they interrupted in order to come to our lecture.

   If the talk is a survey, then you should temper this last advice to suit the occasion. Better to give a survey of a particular aspect of semi-Riemannian geometry than to endeavor to survey the entire subject of geometry. And do suit the talk to the audience. You can survey non-Euclidean geometry for junior/senior mathematics majors, or you can do it for seasoned mathematicians. But you would do it differently for each of these audiences.

2. *Have an attractive title.* A casual observer, seeing the title "Subelliptic estimates for a quasi-degenerate, semilinear partial differential operator satisfying a weak symplectic condition with applications to the hodograph technique of Hörmander," will probably be more tempted to head for a late afternoon beer than to attend your talk. The title "A new attack on a class of nonelliptic equations" conveys the same spirit and is likely to suggest to a broader class of people that the talk may contain something for them.

3. *Prove something.* It leaves a bad taste in everyone's mouth if you talk about a subject but do not get in there and do it. One good strategy is to prove a special case, or work out an example, in some detail; then use this prolegomenon to sketch the key ideas in the proof of the main result.

4. *Structure your talk so that everyone will take something away from it.* Ideally, a member of the audience who is questioned that evening about that day's colloquium should be able to say "The talk was about this" or "The main theorem was that" or "The speaker was relating geometry to combinatorial theory in a new way." If you bear this thought in mind while composing your talk, then it will have a strong, and salubrious, influence on your entire approach to the process.

5. *Be specific.* Heed this advice, both when you are writing and when you are speaking. Nobody wants to listen to an hour of vague fluff. Nobody wants to perceive that you are dodging the main point of the discussion. If you appear to be evasive then, at best, you could make people think that you are sloppy and imprecise; at worst, you could leave people with the impression that you are faking it—indeed that you do not know how to prove these theorems.

### 4.4.8   The Philosophy of a Talk

I once heard a mathematician dedicating a lecture to an eminent person, on the occasion of that man's sixtieth birthday. In brief, the dedicator said "In my country the tradition in lectures, especially with my thesis advisor (whom we are too polite here to name) has been to deal in vague generalities. This man (to whom I am dedicating my remarks) has taught us to present concrete examples, and to work through them completely." The value of showing your audience the inner workings of the material you are presenting cannot be overemphasized. This process helps to draw in students (both young and old), and shows them how the subject works. It also helps to involve those who are not already expert.

6. *Do not be afraid to dream.* I say this cautiously, for I have already warned you not to prevaricate or mislead. But a talk is a different vehicle from a formal piece of writing. Standing before a group and speaking is an

opportunity for you to tell the audience what you tried, what did not work, and what might work in the future. It is absolutely impossible in mathematics to publish a paper that says "Today I woke up and tried to prove the Riemann hypothesis and I failed." In a mathematical *talk*, you can dandle such thoughts before your audience and not only survive, but in fact heighten the audience's appreciation for you and your insights.

7. *Do not be afraid to be informal.* One of the most effective devices that I have seen is for the speaker to say "If we assume these three explicitly stated hypotheses, together with some other technical things that I shall not enunciate, then the following conclusion holds." Often the technical items that are left unspoken are of great interest to the deep-down experts; but to everyone else they would be meaningless, indeed confusing. It takes real insight, and a dash of courage, to be able to say to the audience that you are sloughing over some difficult points. Of course you should never lie; but you may certainly downplay some of the technical points in your subject.

## 4.4.9    How to Present a Proof

These comments also apply when you are presenting a proof. In a specialized seminar, it might be appropriate to slug your way through every technical detail of your argument. In a colloquium, such arcana are virtually never appropriate. If your theorem has any substance at all, then its proof may consist of ten or twenty or more pages of dense argument. It could take a serious reader a week to digest thoroughly the inner workings of your reasoning. Thus it could never work to present the entire theorem, with its proof, in a colloquium talk. Hit the high points, say a word about what you are omitting and oversimplifying. Proud as you are of the cute argument you cooked up for the proof of Sublemma 3.1.5, do not trot it out during your colloquium.

## 4.4.10    Entry Points and Exit Points

8. *Prepare your talk with multiple entry points and multiple exit points.* What does this mean? Rare is the listener who can pay rapt attention for the full space of 50–60 minutes. Many members of your audience will drift in and out. If you say something interesting, then certain people will begin to think their own thoughts, or try to produce a necessary example or lemma. Make it easy for such people to "re-enter" the lecture. Provide several doorways.

Likewise, there is no way to predict how a given talk will go before a given audience. If you are lucky, there will be fortuitous interruptions and serendipitous comments. Time will not be used in just the way that you had planned; you could easily be caught short. Therefore you should create several propitious points at which you can make a gracious exit from the talk. As already noted, a hasty "Egad, I'm out of time" is not a savvy way to finish your colloquium. In any event, do not run overtime—at least not by more than a couple of minutes. First, to do so is rude; second, colloquia are at the end of the day and people have other things to do (such as going home to dinner); third, people simply have no patience for a talk that exceeds the allotted time.

At my university, we had a leading job candidate who, in his ceremonial talk, ran out of time. He gave us a big smile, went to the clock, and pushed the big hand back twenty-five minutes. And then he used them! Suffice it to say that there was no further discussion of his candidacy.

### 4.4.11   Prepare

9. *Prepare, prepare, and prepare some more.* You should have thorough notes before you, but you should rarely refer to them. Your talk should have an edge: you want to be thinking through the ideas with your audience, and you want to be *talking* to the people in the room. You are not giving a recitation to your buddy in the front row. You are not lecturing to the fictitious audience engraved in your frontal lobes. You are talking to the individuals who are breathing the same air as you. Pick them out as you speak; look at them; change your focus and your depth perception as the talk develops. Pace around. Step backward and forward. Use your body and your voice to lure the audience into the talk. Do *not* be a slave to your notes.

Several years ago I watched an eminent mathematician, indeed a Fields Medalist, prepare to give a colloquium on a topic that I personally had seen him lecture on at least four times previously. He had probably lectured on it fifty times in total. I had attended his course on the subject. He *owned* this material; he had created it from whole cloth. He could have given this talk in his sleep. Nonetheless for this, his fifty-first performance, he insisted on sitting quietly in a room for an hour and writing out everything that he was going to say. During the talk, he cast not a glance at his notes. At the end of the talk, he summarily dumped the notes into the trash.

This process made a tremendous impression on me, and I have reminisced frequently about what I observed. *Writing out his talk was his mantra.* He used this process to prepare himself psychologically for the talk. Some people will prepare by just staring off into space and walking, mentally, through the talk. Others will stroll to the student center and buy a cup of coffee. Still others will spend the entire afternoon in the library sweating over the literature, and worrying about questions that someone might ask but in fact never will. It does not matter what you do to psych yourself up for your talk, to guarantee that you are prepared. The main point is to *do something*: find a technique that works for you and use it.

### 4.4.12  Give Proper Credit

10. *Be careful in your talk to give credit where it is due.* Do not give attributions only when your name is involved. In fact most speakers tend to do the opposite. When presenting someone else's theorem, the speaker is careful to write out all the relevant (sur)names in full. When it comes to his own theorem, the speaker just writes something like

> **Theorem:**  [Fu-Isaev-K]
>
> Let $\Omega \subseteq \mathbb{C}^n$ be a smoothly bounded, pseudoconvex domain with noncompact automorphism group . . .

This citation is an example taken from my own life, where Siqi Fu and Alexander Isaev are my collaborators and "K" is yours truly.

Now that I have listed the ten commandments, let me turn to a discussion of general principles. Many technical skills are necessary for giving a good talk. I have already mentioned eye contact and organization.

### 4.4.13  Blackboard Technique

We need to say something about blackboard technique. Blackboards were invented in 1840, and were originally made of slate. For many years they were the lingua franca of teaching. Everybody—from math teachers to English teachers to physical education teachers—used blackboards. Over time, blackboards became less popular and white boards more popular—in part because they do not generate chalk dust, and chalk dust is perceived to be harmful to computers. Nowadays the really popular device for giving a talk is a computer projector. In mathematics, the software to use for preparing the `*.pdf` file for the computer projector is `Beamer`. So let us first say a few words about `Beamer`.

### 4.4.14  `Beamer`

Today the popular method of preparing a math talk is with `Beamer`. `Beamer`, a German product, is a LaTeX package that produces a `*.pdf` file that you project on a screen. It displays your material in slides, and you click through them either with an arrow button on the computer keyboard or with a handheld remote. You should have only about one slide per two to three minutes of speaking.

The advantage of `Beamer` over `PowerPoint` is that, since `Beamer` is a TEX product, it allows you all the mathematical capabilities that you like and need— all the mathematical notation and display, all the fonts, all the superscripts and subscripts, and so forth. A `Beamer` presentation can include graphics, and even animated graphics. It has a customizable color palette. It is really a powerful and flexible tool.

Using a computer projector and projecting your presentation on a screen solves several of the problems associated with blackboards. If you are projecting your material on a raised screen, then you cannot be standing in front of it. And you will probably be using nice fonts, so handwriting is not an issue. In addition, the material will likely be formatted in a nice way. So "blackboard technique" does not come into the picture. On the other hand, presenting your stuff on a screen presents new problems: you might have too much on each screen, you might change screens too quickly, you might have problems going back and forth to refer to earlier material.

I may note that there are times when your main show is on a projector (either slide or computer), but you still may want to write some things on the blackboard. This raises a number of issues. One of these is lighting, because a projector might suggest that the lights be dimmed, while viewing the blackboard may demand more lighting. There are also visibility questions. Just about anyone in the room can see the screen quite clearly. But the screen could obscure the blackboard, or the computer or projector could obscure the blackboard. Finally, you simply must coordinate the two forms of output. Using both media together well requires some planning and some finesse.

### 4.4.15 More on Blackboard Technique

Let me now return to blackboard technique. Even though blackboards are a bit antiquated, there are still many of us who are most comfortable at a blackboard. I still teach *only* using a blackboard. And many others are the same. `Beamer` is great, but it would take a lot of time to prepare a `Beamer` lecture every day. And we older folk are comfortable moving around in front of a blackboard. We are best able to pace ourselves at a blackboard.

A first basic topic is neatness. Even if you are a great expert in your subject, and have a charming and erudite delivery, you will be putting a substantial barrier between yourself and your audience if your writing is an incoherent mess, or if you fill the blackboard with a chaotic barrage of longhand. Learn to write in straight lines, horizontally, from left to right. Write large, and write neatly. Printing is preferable to longhand. Do not put much on each blackboard. Give the audience a chance to copy what you have written before you erase it.

Do not stand in front of what you have written. As you write, read the sentences aloud. Learn to draw your figures accurately and skillfully. Isolate material that will require later reference and *do not erase it*. Plan in advance how you will use the blackboard, so that you can be sure that you will always have room for what you want to write next. Just as the director of a play knows in advance where each actor will be at each moment, and plans every

movement on the stage in considerable detail, so you should plan the moves of
your talk in advance. The audience will grow phenomenally frustrated watching
a forlorn speaker pace back and forth in front of his/her blackboards—for several
minutes!—trying to decide what to erase, or what to save.

Some people solve the blackboard problem by using overhead slides (trans-
parencies) instead. The very act of creating slides in advance addresses virtually
all the issues that I have raised about blackboards in the last paragraphs. Slides,
in the hands of a skilled user, can be a powerful tool. (The blackboard is sometimes inescapable, however, so you should learn to use it.)

If you do use slides, then learn to use them wisely. Each slide should contain
one thought, or one idea. Each slide should contain about six to eight lines,
and should have wide margins. The bottom 2 inches of each slide should be
left blank—because this portion of the slide is often blocked from the vision of
those in the back of the room.

Do not write out complete, long sentences on your slides. Abbreviate wherever possible. I have seen talks in which the speaker simply printed out the text
of a fifty-page TEX document (his/her latest paper) onto transparen-
cies—in 10-point or 12-point type. Moreover, the speaker showed every single
slide to the audience. What a disaster! First, this is far too much material per
slide—and none of it can be read. Second, this is too many slides for a fifty- or
a sixty-minute talk.

### 4.4.16  PowerPoint

Edward Tufte is a notable advocate of good speaking, and of the skillful presentation of graphics to illustrate quantitative information (see [Tuf1], [Tuf2]).
He also, since his retirement from the Statistics Department at Yale University,
gives day-long presentations at hotels on how to give a talk. In these presentations he rails against PowerPoint. In particular, he makes these points about
the software:

- It is used to guide and reassure a presenter, rather than to enlighten the
  audience;

- It has unhelpfully simplistic tables and charts, resulting from the low
  resolution of early computer displays;

- The outliner causes ideas to be arranged in an unnecessarily deep hierarchy, itself subverted by the need to restate the hierarchy on each slide;

- The presentation forces the audience into lockstep linear progression through
  that hierarchy (whereas with handouts, readers could browse and relate
  items at their leisure);

- The technology has poor typography and chart layout, from presenters
  who are poor designers or who use poorly designed templates and default
  settings (in particular, difficulty in using scientific notation);

- The software results in simplistic thinking—from ideas being squashed into bulleted lists; and stories with beginning, middle, and end being turned into a collection of disparate, loosely disguised points—presenting a misleading facade of the objectivity and neutrality that people associate with science, technology, and "bullet points."

See `https://en.wikipedia.org/wiki/Edward_Tufte` for more on this matter.

You may or may not agree with Edward Tufte's vituperation here. But he raises important points that are worth pondering if you are going to use `Beamer` or `PowerPoint` or some other technology to present your talk.

### 4.4.17   Time Management

One of the most important skills that you need to develop, both as a teacher and as a colloquium or seminar speaker, is time management. You need to fit what you have to say into the time allotted. People will be monumentally irritated to watch a mature mathematician spend the last thirty minutes of his/her fifty-minute talk pacing back and forth, scowling at the clock, and declaiming that he/she does not have sufficient time to present his/her thoughts. I have seen many such a speaker act as though it were the audience's fault, or the university's fault, or perhaps his/her host's fault, that he/she did not have more time. What nonsense. The speaker knew when he/she was invited—probably many months before—what the parameters were. Giving a fifty- or a sixty-minute colloquium talk is part of the academic game. Learn to play by the rules.

Perhaps you are saying to yourself—or have said to yourself in the past—"all good and well, but this speech-making stuff is for joke-tellers and hams and showoffs. I am a scholar. I am not an actor." Such a statement is a cop-out (if you will pardon the vernacular). Nobody expects you to be Jerry Seinfeld. Part of a scholar's existence is to communicate—both in writing and in speaking. The thoughts in this section are intended to help you to enhance your abilities with the latter. Giving a talk is a personal affair; you should do it in the fashion that best suits you. But I hope that the ideas presented here will help you to sharpen your wits and your technique.

### 4.4.18   Seminar Talks Versus Colloquium Talks

Just to reiterate one point implicit in the preceding discussion: seminar talks are different from colloquium talks. A seminar is for specialists, including graduate students and faculty. In a seminar you can assume that people are reasonably familiar with the subject at hand and with the relevant literature. You do *not* have to define all your terms. You do *not* have to do a lot of handholding. You can treat some arguments as "obvious and well known." You can get rather technical. And people will appreciate your seminar talk for what it is.

### 4.4.19  The `SmartBoard`

Technology being what it is, people are always coming up with new teaching technologies. One of these is the `SmartBoard`. A `SmartBoard` is an electronic blackboard.

The SmartBoard is an interactive whiteboard produced by the company Smart Technologies. The Smart Technologies Company was founded in 1987, but began producing SmartBoards several years later.

The SmartBoard allows the user to prepare a lecture at home on the computer, bring it to the lecture room on a flash drive, and then plug it into the SmartBoard to get a high-tech presentation. The screens scroll and display very much like `PowerPoint`.[7]

Conversely, one can write on the SmartBoard with a marker and record the lecture as a binary file on a flash drive. Thus it is saved for future use.

The SmartBoard is very smart. It responds to the special pens on the whiteboard, but it also responds to the touch of a finger or other solid object, such as a pointer or stylus.

SmartBoard has its own proprietary software that works with projectors and digital cameras to make the SmartBoard a very versatile and powerful technology.

The user of the SmartBoard can copy, cut, paste, and otherwise manipulate portions of the presentation—very much as he/she would do on a computer screen.

SmartBoard interacts with standard Microsoft software such as PowerPoint, Excel, Word, and AutoCAD.

It is clear to me that preparing a SmartBoard presentation may take more time than preparing a traditional chalkboard presentation. But, once you have the presentation fixed up, then you have it for repeated future use. So it may very well represent time well spent. My personal preference is for a traditional chalkboard, because I can use it skillfully to make the mathematics come alive for the students: They actually get to see *me* doing mathematics step-by-step, just as they will be doing it. It is worth thinking about the merits of the traditional approach.

## 4.5  Your Vita, Your Grant, Your Job, Your Life

### 4.5.1  The Curriculum Vitae

A businessperson has his/her resumé and an academic has his/her Curriculum Vitae (or *Vita* for short). The Vita is your professional history—it should give a quick sketch of who you are, where you were educated, your professional experience, any honors that you have earned, your scholarly accomplishments, and related materials. Usually you will include your publication list with your Curriculum Vitae.

---

[7]Of course one can do something similar with a notebook computer and a projector.

Your Vita should *not* read like this:


> Born on a mountain top in Tennessee.
> Greenest state in the land of the free.
> Raised in the woods so he knew every tree.
> Killed him a b'ar when he was only three.[8]


All quite charming, but a Vita should *never* be in paragraph (or stanza) form. The material should be laid out in a tableau so that the reader can quickly pick out the information he/she needs. Your name should be in boldface at the center top. (I recommend that you use your official name—the one on your birth certificate. Your friends may call you "Goober," but you should save that information for another occasion.) Quickly following should be your date of birth, your educational information, your address and phone numbers and email address, your employment record, key honors earned, and so on. An example of the first page of a Vita appears later in this section.

Your publication list should be a separate section of the Vita. Those who are especially careful separate published works from unpublished (or to-be-published) works and separate items in refereed journals from items in non-refereed publications (such as conference proceedings); books are often listed separately; some people even list class notes they have prepared, or software that they have written (if you are a numerical analyst or a specialist in algorithms, then this last would be essential). Usually the items in a publication list are given in approximate chronological order, although some people use reverse chronological order.

Another section of the Vita lists grants or funding that the person has received over the years. For each grant, you usually list the funding agency, the title and number of the grant, the amount of money in the grant, and the year(s).

Often a Vita will include a section of invited talks or, if you are quite senior, of major invited talks (an hour speaker at a national AMS meeting, or principal speaker at a CBMS conference, or a speaker at the International Congress).

Yet another section will list graduate students (Masters and Ph.D.) that you have directed. Another could list material describing your teaching experience (courses taught, curricula developed, and so forth). Indicate your expertise with computers—either software developed, or courses taught, or other accomplishments.

You may wish to say something about your skill with foreign languages. Have you done any translation work? Are you well traveled? Have you taught in another country?

Finally, some Vitae have a catch-all section with editorial activities, service to professional societies, or anything else that the person writing the Vita thinks may be of interest.

---

[8]From *The Ballad of Davey Crockett,* Walt Disney Productions.

It is common to list on the last page of the Vita the names and contact information for people who are willing to recommend you (for a job or an encomium or something else). These should of course be people whom you have contacted and asked permission to use their names.

## 4.5.2   Overstating the Case

Your Vita is no place to be humble. This document is the *gestalt* that you present to the world. Certainly do not prevaricate—or even exaggerate—but be sure to tell the reader everything that you want him/her to know about yourself.

At the risk of sounding preachy, let me expand a bit on one of the points in the last paragraph. When preparing the Vita, we all want to present ourselves in the best possible light. There is a tendency to dress things up—beyond what is strictly kosher. Perhaps you did not complete that French course—but you ate quiche Lorraine once—so you write that you speak French. The letter from the journal to which you submitted your latest paper says "if you make the following twelve changes then the referee will have another look at it," and you list the paper as accepted. The NSF tells you that you are on the "maybe" list for a grant, and you put on the Vita that you have a grant. People who have made these slips are not liars; they are just trying too hard. Strive to avoid such exaggerations. Most departments check facts carefully. Many schools only believe in publications that have appeared, and for which there is a *bona fide* reprint (many schools have been burned once too often in the past). If the Funded Projects office at your school does not have the letter from the NSF, then your grant does not exist. Worse, if you make such claims in your Vita and the claims do not wash with the people evaluating your case, then the situation will weigh against you. My advice is to be extra careful.

SAMPLE VITA

# CURRICULUM VITAE
for
### Clemson Ataturk Kadiddlehopper

**Date of Birth:** March 15, 1947

**Home Address:** 17 Poverty Row, Faculty Ghetto, Iowa 50011

**Current Academic Affiliation:** Department of Mathematics
Walmart A&M, Sam's Clubville, Iowa 50012

| | |
|---|---|
| **Telephone:** | (515) 294-6021  (office) |
| | (515) 373-3286  (home) |
| | (515) 294-6047  (FAX) |
| **email address:** | cak@math.sam.edu |
| **Graduate Education:** | *Ph.D., Mathematics* |
| | Montana Institute for the Tall, 1974 |
| | Thesis directed by Charles Ulmont Farley |
| | *M.S., Mathematics* |
| | Frisbee State University, 1971 |
| **Undergraduate Work:** | *B.A., Mathematics* |
| | Joe's Bar and University, 1969 |
| **Academic Positions:** | Assistant Professor, College of the Yodeling |
| | Yuppie, 1974–1979 |
| | Associate Professor, Steland Lanford |
| | University, 1979–1988 |
| | Professor, Walmart A&M, 1988–present |

**Honors:** Neural Sediment Fibration Graduate Fellow, 1971–1974
Visiting Professor, Callipygean Institute of Tectonics, 1977
Shinola Fellow, College of Good Hair, 1979
Visiting Professor, Upper College of
     Lower Academics, 1980
Visiting Professor, University of Basic Bourgeoisie, 1986
Visiting Professor, Hahvahd University, 1986
Honorary Lecturer, Crab Louie College, 1987

### 4.5.3   Tailoring Your Vita

Now let us return to matters prosaic. You must tailor your Vita to the circumstances. I have been teaching for 43 years. Thus it would be crazy for me to list every course that I have ever taught. It would make more sense for my Vita to list courses that I have created, or textbooks that I have written. On the other hand, if you are just starting out in the profession, then you should indicate the depth and range of your teaching experience and certainly indicate your facility with computers, both in the classroom and outside it. A beginner will probably have directed few if any graduate students. Not a problem, since such activity is not expected. Do, however, be complete in describing your other activities.

## 4.5.4   The Funding Picture

Funding is available for many different types of activities that a mathematician might undertake. These range from quite specific, goal-oriented projects that are funded by industry all the way to grants from the NSF (National Science Foundation) to encourage pure research in abstract mathematics. There is also funding from the Department of Defense, from DARPA (the Defense Advanced Research Projects Agency, an agency of the Department of Defense), from NASA (the National Aeronautics and Space Administration), from NIH (the National Institutes of Health), from DOE (the Department of Energy), from NSA (National Security Agency), from the Simons Foundation, and from many other sources as well. Granting agencies such as the NSF have considerable funds to encourage work on the mathematics curriculum—from developing new ways to teach calculus to revising substantial blocks of undergraduate mathematics education.

Given the range of activities that granting agencies are willing to fund, and given the variety of different potential sources of grants, I could discuss grantsmanship at length. I shall content myself here with a few general precepts that should apply to virtually any grant application that you may write.

## 4.5.5   Preparation for Writing a Proposal

Read the prospectus for the program to which you are applying. Doing so, you will learn what the program is looking for, what particulars should be itemized in the proposal, what page limits will be enforced, and when the proposal is due. Learn about what type fonts are acceptable, what margins the pages should have, how long the Curriculum Vitae portion of the proposal should be, how long the references section should be, how many pages should address previous work, how many pages should address new work ... *and so forth*. Grant proposal writing is not a free form activity. Get the rules straight before you begin.

The main issue in the air when your grant proposal is being evaluated is your credibility: *can* you do the work being proposed, and *will* you do the work being proposed? Given your stature, your abilities, and your track record, is it clear that you can work on these problems (be they research or education)?

Can you solve them or make progress on them? Are you capable of evaluating your own progress? Finally, can a case be made that you are *the right person* to work on this project? Or will the work be done as a matter of course by others (if indeed it is worth doing at all)?

### 4.5.6 Your Credibility

You must walk a delicate line here. On the one hand, you want to make it clear to the potential granting agency that you know this subject inside and out, that you know the existing literature, and that you have a good program for proceeding. You want to demonstrate that you are already engaged in some version of the proposed activities. On the other hand, you do not want to make it sound as though you have already solved the proposed problems. You also want to give the strong impression that you are working on substantial problems of real significance; these should be problems for which even partial results will be of interest. But it should be plausible that you are up to the task. In particular, if you propose to prove the Riemann hypothesis, then you will have a difficult time making your case. After all, many of the bigshots are working on this problem; if they cannot make inroads, then how will you?

Generally speaking, granting agencies will not provide funds to help you to learn something new, or to retool. Thus you should make a case that you are already engaged in the proposed project, that you have a grip on it, and that you have a viable program. If those considerations entail your learning something about nonlinear elliptic PDEs, then by all means you should say so. But a grant proposal that reads (in effect) "I'm tired of studying finite groups so now I'm going to do symplectic geometry" just will not fly.

As already noted, you must prove that you are up on the relevant field— not just what is in the published literature but what is available in preprint and other tentative form. For the most part, grants are refereed by your peers. These will be peers who are on the cutting edge. They will judge you by their own standard—the standard by which they themselves would expect to be judged.

### 4.5.7 Page Limits

When writing a grant proposal, you must walk another delicate line. You will usually have a page limit. You simply cannot go on at length, or have extensive digressions, or have verbose introductions or chatty conclusions. But you must make the proposal as easy to read, and as self-contained, as possible. If your proposal engages the referee's interest, and teaches him/her something, and does not force him/her to keep running to the library to figure out what you are talking about, then you will be at a real advantage. If, instead, your prose is a bore and the referee has to slug his/her way through it, then your proposal is likely to be penalized.

Do not be afraid to telephone the granting agency to which you are applying and to talk to the program officer. Many grant programs, and many program officers, encourage this activity. By talking to a program officer, you can better focus and tailor your proposal to the goals of the intended program, and you will not waste the program's time with a proposal perceived to be completely off the wall.

## 4.5.8   Proposals in Math Education

Proposals in mathematics education and curriculum often require a section on "self-evaluation" and a section on "dissemination." You should not (in the self-evaluation portion of your proposal) say "We'll see how happy the students are at the end by distributing teaching evaluations." You also should not (in the dissemination portion of your proposal) say "I'll go to conferences and talk about this stuff with my buddies." I have seen both of these in serious proposals, and they do not work. Both approaches are too facile, and show no imagination and no effort. Good self-evaluation programs often involve motivational psychology experts from your institution's School of Education, tracking of students after they leave the experimental program, exit interviews, and many other devices. Good dissemination programs often involve writing a textbook for publication, creating a newsletter, setting up a web page, organizing workshops, and so forth. I am not necessarily advocating any of these devices. I am merely explaining how the world works.

## 4.5.9   Self-Evaluation and Dissemination

Self-evaluation and dissemination play an implicit role in a research proposal as well. Your report on previous work will give an indication of your ability to evaluate your own progress. The scientist who says "In the last five years I tried a lot of things but nothing panned out" shows both bad judgment and an inability to learn from his/her work. The dissemination aspect of a research proposal is reflected in your publication record, your invitations to speak at conferences or colloquia, and your collaborative activities. NSF proposals now require a section on dissemination and impact.

Before you submit your proposal, run it through a spell-checker. Check and recheck the grammar. Show it to a senior colleague. Proofread it more times than you think could possibly be necessary, and then proofread it once more. The reviewer will be phenomenally irritated to read a proposal that appears to have been prepared hastily or sloppily. Be sure that you have included all the relevant references. It is quite likely that some of your reviewers will be expert in your subject and will expect their work to be cited. Do not disappoint!

Your proposal should be as slick as glass. It should be a pleasure to read, and it should get the reviewer excited about and interested in what you are proposing to do.

### 4.5.10 Applying for a Job

At one time or another, most of us will have to apply for a job. Let me first say a few words about applying for an academic job.

When you apply for a job at a college or university, you send in your Vita (discussed above) and a cover letter. The cover letter should be brief (well under a page); it should identify you, your present position, the type of position you seek, and your areas of interest. No more. A cover letter with a multitude of exclamation points, shamelessly extolling your virtues as a teacher and your bonhomie as a colleague, is highly inappropriate. A sample cover letter appears later in the section.

### 4.5.11 MathJobs

I would be remiss not to note that the AMS (American Mathematical Society) has created an online utility called mathjobs. This is a device for applying for a job. If you are a candidate, then you arrange for all your materials (your letters of recommendation, your Teaching Statement, your Research Statement, your Vita, your cover letter, and so forth) to be uploaded to the mathjobs website. Then any school interested in you has complete access to all your materials. There are many advantages to this system: **(i)** all your materials are in one place, **(ii)** the materials cannot be lost or misplaced, **(iii)** several people can view your materials at once. It is safe to say that mathjobs has become a cornerstone of the job marketplace. Not every school participates in mathjobs, but a great many do. If you are a senior person applying for a position, then using mathjobs may be inappropriate. The school to which you are applying will let you know.

### 4.5.12 Things to Include in Your Application

If you are a beginner in the profession, with a short publication record, then you might include some of your preprints with the job application (these could be in electronic form). Your Vita should also list your "references" or "recommenders." This point is vital, and many job applicants overlook it. Before applying for any job you should approach three or four people (for a senior job it could be six or eight, or even more) and ask whether they are willing to write in support of your application. Ideally, these should be prominent people in your field, whose names will be recognized by those evaluating your dossier.[9] Once you induce a group of such people to agree to write,[10] then include their names, business addresses, business phone numbers, email addresses, and fax numbers in a section of your Vita called "References." These days—especially if you are a job candidate just a few years past the Ph.D.—you should have one

---

[9]If you are a very senior candidate for a position, such as I was for my most recent job, then you do not contact the letter writers. The hiring institution will do so.

[10]The notion of the candidate *asking* people to write for him/her seems to be a peculiarly American custom. In many countries—especially in Europe—the hiring institution does all the solicitation of letters.

or two letters from people who can say something specific and positive about your teaching. Their names should be listed in your References section as well. It is now commonplace for job candidates in the United States to include the "AMS Standard Cover Sheet" in the dossier; this form may be obtained from most issues of the *Notices of the AMS*.[11]

Some job candidates arrange to have a sample of their teaching evaluations, or passages from their present institution's Teacher Assessment Book (such a book is often published by the campus student organization) to be included in their dossier. Such an inclusion can help to lift the dossier out of the ordinary, and will add substance to the letters that praise the candidate's teaching abilities.

If you use your imagination, you can probably think of all sorts of things that might be included in your dossier in candidacy for a job. I recommend that you consider each one cautiously. People on hiring committees these days often must wrestle with 500 job applications in a season. A big, fat dossier will just turn off a weary committee member.[12] So do not leave anything important out of your dossier, but think carefully about what you do include.

If you make your application in the manner described in the preceding paragraphs, then a school to which you apply will know just how to process your paperwork. Once it has your cover letter and Vita, it can start a file on you. Then it has a place to put the letters of recommendation as they come in. And, because you have included a list of references, the school will know when your dossier is complete.

### 4.5.13   The Job Visit

If your application is for any position beyond a beginning lectureship, and if you make the "short list," then you will likely be invited to give a talk and to meet your potential future colleagues. Let me not mince words: this is a make-or-break situation. Dress well (not as though you were entering a ballroom-dancing contest, but rather as though you are taking the situation seriously). Give a polished, well prepared talk (see Section 4.4 on how to give a talk). Think in advance about some of the topics of conversation that may come up when you meet your new colleagues. Be prepared to describe your research conversationally to a small group of nonexperts; be able to say in five or ten

---

[11]Many people, especially young job candidates, include in their job application a one- or two-page statement describing their research program; many young people also include a "statement of teaching philosophy." The first of these can be quite helpful to a nonexpert who is endeavoring to evaluate the dossier: a good research statement can at least guide such a reviewer to an appropriate expert colleague who can comment on the case in detail. I dare not say whether a "teaching philosophy" statement has any real value; there is little grass growing in this subject area, and you will have trouble finding anything interesting or original to say. On the other hand, many schools *require* a statement of teaching philosophy; in such a circumstance you must do your best to write something thoughtful and thought provoking.

[12]It is quite common these days for a high school student applying to college to include a video, or a videotape, in his/her application. Such an item would be quite inappropriate in a professional job application.

minutes what you do, how it fits into the firmament, who are some of the experts, what are some of the big questions.

*Be prepared to say who in this new department has interests in common with yourself; with whom you think you might talk mathematics; who might become your collaborator.* Do not underestimate the significance of this circle of questions. Do *not* say "Oh, I talk to everybody; I'm the Leonardo da Vinci of modern mathematics." Such a statement is not credible; utter it and you will surely send yourself plummeting to the cellar of the short list.

### 4.5.14 Teaching Topics

Think over your ideas about teaching, about the teaching reform movement, about teaching with calculators or computers, about teaching students in interactive groups, about teaching reform (see [Kr2]), and about any other topics that may arise. Some schools have special problems connected with the teaching of large lectures; be prepared to share your views on that topic. Other schools have special tutorials for calculus students; be prepared to chat about that topic as well. Some schools like to conduct a formal interview, with a few of the senior faculty asking you direct questions about your research, your teaching, your attitudes about curriculum and reform, about teacher/student rapport, or anything else that is in the air at the time. It makes a dreadful impression if you are inarticulate, do not seem to know your own mind, or simply have not given any thought to these matters. I am not advocating that you go to your ceremonial job interview with a sheaf of notes *in your hand;* I am instead suggesting that you go with a few note cards *in your head.*

### 4.5.15 The Job Offer

When a school decides to offer you a job, the Chair will usually telephone you, or send you an email message followed by a phone call. At that time he/she may discuss salary, teaching load, computer equipment, startup funds, health insurance, the retirement annuity, or other perquisites. You may wish to take the opportunity to ask about these or about other concerns.

Important information about you can be lost in the Vita—especially if your Vita is long. If you are a graduate student applying for a first job, and if you have won a teaching award for "Best TA," then certainly mention that encomium in your cover letter. If you are a few years from the Ph.D. and the holder of a Sloan Fellowship, or an NSF Postdoc, then you should mention these honors in your cover letter. You do not want your cover letter to look like a flyer from your local supermarket, but you want it quickly to lead the reader to your strong points.

SAMPLE COVER LETTER

_____

November 22, 2018

David P. Davids, Chair
Department of Mathematics
Little Sisters of the Swamp College
Sanctuary, Oklahoma 23094

Dear Professor Davids:

I wish to apply for a faculty position, at or near tenure, in your department. I received the Ph.D. in Mathematics in 2013. I am a geometric analyst, with specializations in complex function theory, harmonic analysis, and partial differential equations. My Vita is enclosed. It includes my list of references. I currently hold the position of Instructor of Mathematics at Brouhaha Subnormal School in Wichita Falls.

Please note that my research is supported by a grant from the Normative Sodality Agency. I am also a recipient of the Mudville Distinguished Teaching Award. I have strengths in research, teaching, and curriculum.

I look forward to hearing from you.

Sincerely,

Robert Q. Roberts
Instructor of Mathematics
College of the Yodeling Yuppie

_____

### 4.5.16 Jobs in the Private Sector

If you are applying for a job in the private sector—say at Texas Instruments, or AT&T, or Aerospace Corporation, or Microsoft—then the application process may be a bit different from the process in an academic setting. An industrial organization is probably not interested in letters of recommendation from the Université de Paris, nor in binary comparisons with famous young algebraic geometers. A business resumé is different from an academic Vita. Go to `Amazon` and purchase a book on how to write an effective resumé (see, for example, [Ad3]), how to write a cover letter (see [Ad1]), and how to apply for a job (see [Ad2]). Although working in industry certainly will involve communication skills, it probably will not involve much classroom teaching. The interview for an industrial job will likely be even more crucial than the interview for an academic position. Consult acquaintances who have been through the process so that you can be well prepared.

### 4.5.17 Writing That Has an Impact

In the abstract, the rewards for good writing may seem far off and vague; instead you can see clearly how a well-written Vita or grant proposal could lead to just deserts. Your Vita is a tool for helping you to find employment, or a promotion, or to achieve some other goal. Your grant proposal is a way to seek funding. I have also discussed how to find a job. The principles of good writing described in other parts of this book apply just as decisively to these practical matters: express yourself directly, cogently, and briefly; do not show off; know what you are talking about; and (paraphrasing actor Jimmy Cagney) plant both feet on the ground and tell the truth.

### 4.5.18 Unusual CVs

I have seen many a Vita in my time. One of these contained a page entitled "Cities Beginning with the Letter 'Q' in which I Have Spoken Fewer than Five Times." Another listed an uncompleted mystery novel. Yet another listed forty (count 'em) collaborative papers that were incomplete and in progress. One Vita by Mathematician $X$ listed poetry, both published and unpublished, that was written to $X$, by $X$, about $X$. Another Vita listed the subject's (not very happy) marital history. Your Vita is a business document. This piece of paper is a précis of your professional life. Think carefully about what you put into it and how you organize it.

   Your grant proposal is a manifestation of your professional values, what you are all about, and what you are trying to do. As you develop it, read it with the eyes of your potentially most critical reviewer.

   Here we are discussing writing with immediate impact, and with a direct effect on your life. This is writing that you wish to succeed because it must. Even more than in your other writing, you will want to strive to make each word count, and to force each sentence to say precisely what is intended. The critical skills discussed in this book should help you in these tasks.

## 4.6   Electronic Mail

### 4.6.1   Communicating Electronically

Electronic mail (or email for short) is an important part of professional life.[13] The technology of email enables us to carry on extended conversations with people all over the globe. We can engage in topic-specific discussion groups, conduct business, have professional collaborations, develop friendships, and even have fights via the Internet. Perhaps more significantly, we can conduct mathematical collaborations with people 10,000 miles away, in some cases with people whom we have never even met. You may actually (though I encourage you to exercise this option with discretion) send an email blind to a professor at M.I.T. and say "Hello, I'm so and so. Do you know the answer to the following question?" I have occasionally engaged in this speculative activity and, more often than not, I have received a useful answer.

Several years ago I was writing a series of papers with two collaborators, one of whom is usually in Los Angeles and the other usually in Canberra, Australia. At one point, one of us spent a leave in Berkeley, another took a leave in Wuppertal, Germany, and the third changed jobs. We did not miss a beat, because email is oblivious to these moves. Marshall McLuhan [McL2] died too soon: the global village is finally here in spades.

### 4.6.2   The Nature of Collaboration

G. H. Hardy and J. E. Littlewood carried on what is by now the most famous, and certainly the most prolific, mathematical collaboration in history. Usually in two different locales (one in Cambridge, the other Oxford), they conducted their collaboration by regular post (now known as "snail mail"). Their hard and fast rule was that, if one of them received a letter from the other, he was under no obligation to open it—right away or at any time. Many a letter was thrown into a pile, not to be read then or perhaps ever; this to guarantee that the recipient could think his own thoughts, and not be interrupted. To my mind, email is a bit different: it gets right in your face, once or several times a day. Once you have determined (perhaps by looking at the Subject Line) what a particular email message is and whence it came, then you are looking at it. The temptation is to read it. As McLuhan taught us [McL1], "the medium is the massage."

For many purposes, communicating via email is preferable to communicating by telephone. For an email message has the immediacy of a telephone call without any of the hassle of playing "telephone tag," talking to voice mail, or patiently explaining your quest to a secretary. Many of us find that we send

---

[13]In personal situations, people often use text messages instead.

more email than we do letters, and we use email more often than we use the telephone.

### 4.6.3 Email Etiquette

Because the use of email has become so prevalent, we must all learn some basic etiquette of the email system. As with many other activities in life, email is something that we can benefit from if we give it just a few moments of reflection. Here are some notable points about email:

- Be sure that your email messages go out with a complete header. This should include an informative Subject Line. If people use Microsoft `OutLook` or `gmail` to download and sort their email s, then the Subject Line is essential to the process.

- These days most email is sent using `gmail`, either with the web-based interface or else with apps available on Android or Apple iOS devices or Apple's Mail program. All these utilities have attractive features to make the email process both simple and easy.

- I often receive email messages that say (*in toto*) "Yes, I agree with you completely" or "Right on" or "There you go again!" I love fan mail as well as the next person, but I often cannot tell what such email is about. Do yourself and your correspondent a favor and either **(i)** include the email message to which you are responding in your reply or **(ii)** at least include a sentence or two indicating to what you are responding.

- Sign your email message with your full name. Signing off with "See you later, alligator" or "That's all, Folks" is momentarily amusing, but it often forces your recipient to search the header of the message to determine whose pear-shaped tones he/she is reading. Such a search is sometimes frustrating, and irksome to boot.

What is best is to have a several-line signature (see below) that tells the recipient in detail just who you are.

The best possible "signature" to an email message is a signature block like this:

```
*********************************************************
* Steven G. Krantz  (314) 935-6712  FAX (314) 935-6839*
* Department of Mathematics, Campus Box 1146         *
* Washington University in St. Louis                 *
* St. Louis, Missouri 63130-4899  sk@math.wustl.edu  *
*********************************************************
```

`Gmail` and other email clients make it straightforward to formulate and install such a message. In many cases it is automatic.

- I have arranged for all my university email to be forwarded to my `gmail` (i.e., Google mail) account. I am happy to say that `gmail` is a powerful system that will allow one to create a signature and to customize your email s in a number of useful and attractive ways. I note that `gmail` has a terrific spam filter. And `gmail` allows you to send attachments of any size. It is a great system.

- The good news about email is that it is a lot like conversation. It is spontaneous, natural, and candid. The bad news about email is that it is a lot like conversation—without the give and take of an interlocutor. Thus we are tempted to type away madly, at high speed, having no care for corrections or proofreading. This is a big mistake.

### 4.6.4   Important Emails

- Proofread each email message before it goes out. If the message is important, then proofread it several times. In any event, learn to use the editor on your system and *use it*. Correct misspellings (many an email editor is equipped with a spell-checker) and misstatements. Clean up your English. Some email messages that you send will have the permanence of a hard copy written letter. Send something that will reflect well on you.

- In fact, when I am writing something of great importance, I compose it on my home computer—using the text editor with which I am most familiar (see Section 6.3 for a discussion of text editors). I do this in part for psychological reasons. When I compose on my home computer, I do not worry about the system hanging or going down; I do not worry about taking a break and being thrown off the system; and I am using a writing environment with which I am thoroughly conversant. I can use my spell-checker, my optical-disk dictionary and thesaurus, and other familiar resources to put the document in precisely the form that I wish. I also can sleep on the matter before I send the document.

    The next morning, I bring the document to work on a flash drive (or perhaps I uploaded it to `DropBox`), upload it to the system (see Section 6.8), and then pull it into an email message using operating system commands. You can also use `Windows` commands (like `cut` and `paste`) to effect these moves. This methodology is a valuable tool.

- Implicit in the preceding discussion is a major liability of email. Too easily can you write something in haste in the email environment and then just send it off—it only requires a key stroke or two!—and then it is gone. You cannot retrieve it (Well, sometimes you can. But it is a little tricky, and you can only do it for a brief time window). Unfortunately you may decide, a few minutes later, that you wish you had *not* sent that hot-headed email. You really need to discipline yourself to avoid shooting yourself in the foot with email.

## 4.6.5   Fighting over Email

I once had a rather significant fight with another mathematician. He wrote me a letter taking me to task for something that I had done. The writer was a friend, and someone whom I respected.

Fortunately, this event occurred in the days before email. I wrote a hasty and heated response (in hard copy, for that was all that we had at the time) telling this person that he was misguided and mean-spirited, and just plain wrong. I dropped it in the department's outgoing mail tray. An hour or two later, I pulled the letter from the mail (I had been stewing about it all the while), and penned a milder version of the heated letter. This revision process repeated itself throughout the day. By the end of the day, I had put in the mail a letter of apology, acknowledging my error and thanking my correspondent for calling it to my attention. And further promising various ways in which I would fix the situation. I have always been happy for this outcome. With email the story would have ended differently, and badly.

## 4.6.6   More Email Tips

- In the same spirit as the last paragraph, be careful not to reply to email s too quickly. It is just too easy to dash off a thoughtless reply that you will regret later. Pause for a moment, take a sip of coffee, think about what you want to say.

- Try to keep your email messages brief. Of course I realize there are times when you are circulating a report or writing a detailed formal analysis of some situation; in such circumstances, it may be appropriate to go on at some length. But, most of the time, when writing email, you are sending a memo. Thus make it quick. Often, on the computer, we tend to do things just because we can. Writing an email message is a lot like talking, but without the reality check of having someone interrupt you from time to time. Thus you must show some good sense: say what you have to say, say it cogently and completely and *concisely*, and then cease.

- Unlike conversation, email is toneless. It has no inflection (as with a voice). It can easily be misconstrued. You can, in an email, inadvertently insult someone. Your intentions can be misconstrued. If you—even unintentionally—make snarky comments, then they could get forwarded or they could come back on you later.

## 4.6.7   The Form of an Email

- You should think carefully about how to construct an email that you are writing. If it is just a note to an old friend, then you can dash off whatever pops into your head without fear of reprisal. But if it is an email to an important and busy person, then your first sentence should say right off the bat what your main point is. The next few sentences should flesh out this main point. Then the details can follow.

- A *really* busy person most likely will not read your entire email, so you want to get the main message to him/her at the very outset. The recipient can use his/her own judgment as to how far to proceed in the email.

- You may also find it useful to highlight certain passages in your email with another color. The utility `gmail` makes it very easy to do this. Using other fonts can also be helpful. In addition, you can render some parts of your message in larger type. These are all different ways to draw attention to your main points. 1

## 4.6.8   Forwarding Email

- You can easily forward any email message that you receive to anyone that you like. I am astonished at the extent to which this power is misused. When you receive a hard copy letter of recommendation in the mail—for a tenure case, say—you probably do not make 50 photocopies of the letter and send them off to 50 different mathematicians. First, such an action would be rude; second, it could have legal repercussions. For a written letter, the sender owns the contents and the recipient owns the piece of paper and that particular *form* of its contents (this is the law). What I am discussing here is not so much the law as common sense and common courtesy.

   People forward email all over the place, with hardly a thought for the consequences. The courteous thing to do is to ask the author before you forward anything. Many people send me email messages that say "Please delete this message after you have read it" (the implicit message here is "Don't forward this to anyone!"). I am always punctilious about adhering to such a request, and I hope that others are similarly considerate of my requests for discretion.

### 4.6.9   Other Features of Email

- Email has certain troublesome features like `bcc`. This utility allows you to send blind carbon copies of your email to certain people—so that the main recipient, the person whose name is in the header—cannot see who is getting the copies. I never use `bcc` myself. I think it is dishonest, and I have no need for it. But many people use it regularly.

- Another useful feature of email is the "Reply All" button. This allows you to reply not just to the sender, but to all the recipients of the email. Of course you must use such a utility with caution: there are some replies that you *want* everyone to see; others not.

- Electronic mail is more secure than it used to be. All `gmail` messages are encrypted, for instance.

Electronic mail, or email, is a marvelous tool. It has affected the mathematical infrastructure, and has altered the way that many of us collaborate and communicate. If each of us would exercise just a little email etiquette, then the annoyances attendant to email would be minimized.

## 4.7   Copyright

### 4.7.1   Protecting Your Work

Copyright is a legal right created by the law of a country that grants the creator of an original work exclusive rights for its use and distribution. This is usually only for a limited time. The exclusive rights are not absolute but restricted by limitations and exceptions to copyright law, including fair use. A major limitation on copyright is that copyright protects only the original expression of ideas, and not the underlying ideas themselves.[14]

Copyright is a form of intellectual property, applicable to certain forms of creative work. Some, but not all, jurisdictions require "fixing" copyrighted works in a tangible form. It is often shared among multiple authors, each of whom holds a set of rights to use or license the work, and who are commonly referred to as rightsholders. These rights frequently include reproduction, control over derivative works, distribution, public performance, and "moral rights" such as attribution.

Copyrights are considered territorial rights, which means that they do not extend beyond the territory of a specific jurisdiction. While many aspects of national copyright laws have been standardized through international copyright agreements, copyright laws vary by country.

---

[14]It is a fact that you cannot copyright a title. This is why there are so many books entitled *Calculus with Analytic Geometry*. Even Michael Crichton could not copyright the title *Jurassic Park*. But he *could* copyright the concept of Jurassic Park.

## 4.7.2 Your Writing Is Copyrighted to You

It is a fact that, as soon as you write something, it is copyrighted to you. You, as a writer, may find this fact comforting. When you write a book for a publisher, or a paper for a journal, then it is common (although not always mandatory) for you to sign the copyright over to the publisher. This makes both logical sense and legal sense since the copyright is yours to either keep or to assign to someone else. Many publishers really only want the distribution rights to your work, and such publishers may not insist on having the copyright.

## 4.7.3 Duration of the Copyright

Typically, the duration of a copyright spans the author's life plus 50 to 100 years (copyright typically expires 50 to 100 years after the author dies, depending on the jurisdiction). Some countries require certain copyright formalities to establish copyright, but most recognize copyright in any completed work, without formal registration. Generally, copyright is enforced as a civil matter, though some jurisdictions do apply criminal sanctions.

Most jurisdictions recognize copyright limitations, allowing "fair" exceptions to the creator's exclusivity of copyright and giving users certain rights. The development of digital media and computer network technologies have prompted reinterpretation of these exceptions, introduced new difficulties in enforcing copyright, and inspired additional challenges to copyright law's philosophic basis. Simultaneously, businesses with great economic dependence upon copyright, such as the music business, have advocated the extension and expansion of copyright and sought additional legal and technological enforcement.

The fact is that copyright law is quite complicated, and international copyright law is even more complicated. We cannot delve into the details here. The website

```
https://www.copyright.gov/help/faq/faq-general.html#patent
```

has some good basic information about copyright.

## 4.7.4 Creative Commons

A Creative Commons (CC) license is one of several public copyright licenses that enable the free distribution of an otherwise copyrighted work. A CC license is used when an author wants to give people the right to share, use, and build upon a work that they have created. CC provides an author flexibility (for example, he/she might choose to allow only non-commercial uses of his/her own work) and protects the people who use or redistribute an author's work from concerns of copyright infringement as long as they abide by the conditions that are specified in the license by which the author distributes the work.

CC licensed music is available through several outlets such as `SoundCloud`, and is available for use in video and music remixing.

There are several types of CC licenses. The licenses differ by several combinations that condition the terms of distribution. They were initially released on December 16, 2002 by Creative Commons, a U.S. non-profit corporation founded in 2001. There have also been four versions of the suite of licenses, numbered 1.0 through 4.0. As of 2016, the 4.0 license suite is the most current.[15]

# 4.8 If You Are Not a Native English Speaker

## 4.8.1 English as the Default Language

On the one hand, it is rapidly becoming the case that English is the default language for writing in mathematics. If you want your ideas to be read around the world, you write in English. French and German and Japanese are fine, but they do not have the universality of English.

But many of us are not native English speakers. Our English may be serviceable, but it is not perfect. We may need some help to get our prose up to snuff for publication. What are the options?

## 4.8.2 Help with English

First, many publishers can offer help with English. They have staff professionals who are skilled at helping non-native speakers sharpen their prose. They know enough about other languages that they know which bugs to look for and how to fix them. Do be sure to ask your publisher whether they can provide such assistance if you need it.

Second, there are professional private writing coaches. Of course they are for hire, so you will have to find the resources to remunerate them. But the expense will probably be worth it. And you may, in the process, gain a valuable ally for future writing projects.

It is also possible that you can get one of your students—either an undergraduate or a graduate student—who is a native English speaker to help you with your writing. The student will likely be thrilled to be asked to assist with a scholarly task. And you will enjoy working with a student on a worthwhile project.

Finally, you may have a generous and friendly colleague who is willing to give some time to helping you with your writing. After all, this is in part why we have colleagues. And you will feel quite comfortable working side-by-side with a colleague of your own age and with similar training.

## 4.8.3 Gifted Writers

As time goes on, if you work at it, your English will become better and better. Some of the best writers in English that I know are not native English speakers.

---

[15]In October 2014, the Open Knowledge Foundation approved the Creative Commons `CC BY`, `CC BY-SA`, and `CC0` licenses as conformant with the "Open Definition" for content and data.

A good example is Benoit Mandelbrot [France], who was an extraordinarily gifted writer. Indeed a good deal of his success can be attributed to his writing ability. Another example is Elias M. Stein [Belgium] at Princeton. His analysis books have been enormously influential, and are widely read with great pleasure. A third example is Paul Halmos [Hungary]. Paul became one of the pre-eminent mathematical writers of the twentieth century. He was also a terrific editor and teacher. He certainly helped me a lot with learning to write, and he played a pivotal role vis a vis some of my most important publications.

# Chapter 5

# Books

*Some books are to be tasted, others to be swallowed, and some few to be chewed and digested.*

Francis Bacon
*Essays* [1625], Of Studies

*No man but a blockhead ever wrote except for money.*

Samuel Johnson
quoted in Boswell's *Life of Samuel Johnson*

*I never think at all when I write*
*nobody can do two things at the same time*
*and do them both well.*

Don Marquis

*A writer and nothing else is a man alone in a room with the English language, trying to get human feelings right.*

John K. Hutchens

*The writer who loses his self-doubt, who gives way as he grows old to a sudden euphoria, to prolixity, should stop writing immediately: the time has come for him to lay aside his pen.*

Colette

*You can't polish cow chips.*

paraphrased from Lyndon Johnson

## 5.1   What Constitutes a Good Book?

### 5.1.1   Writing a Good Book

Mathematics books are written all the time. Go to the library and pull one at random off the shelf. Looks pristine, does it not? Or perhaps only the first fifty pages show signs of reading. Many an author lavishes all his/her enthusiasm and creativity and energy on the first part of his/her book; he/she then runs out of steam for the remainder. Unfortunately, it is the reader who suffers the consequences.

Writing a good book requires more effort than many authors are willing to give to the task. Writing a good *mathematics* book requires special insights and skills. In my view, the hard work is worth it. When you write a good mathematics paper, it is only read by a small group of people. But write a good book and a lot of people will see it. The book is a way of planting your flag, of putting your stamp on the subject, of sharing with the world the fruits of your hard labor.

My advice is not to consider writing a book until you have tenure and are planted somewhere. The task is just too time consuming, and is often not construed as a positive contribution toward the tenure decision. Put differently, and a bit simplistically, the view of the world is that an Assistant Professor should be writing research papers and becoming established in the research community.[1] Once you have done that, and achieved tenure status, then you have the leisure to consider other pursuits.

Now let us consider what makes for a good book. First, and foremost, you must have something to say. If you are only repeating, or paraphrasing, what has been said before, then you are contributing nothing to the subject. Second, you must have a plan for saying it. The best method for writing a book is to immerse yourself thoroughly in the subject. The book itself becomes your "world" for a couple of years. A place to begin is to write a detailed outline of the book. Begin by writing chapter headings. Then fill in some section headings. After a while, the juices begin to flow and you will find that you cannot write fast enough to keep up with the outline developing in your head.

### 5.1.2   How the Book Unfolds

Once the book outline is written, it should be emblazoned on your frontal lobes. Carry it with you (in your head) all day long. I find, when writing, that I am constantly jotting down thoughts or topics or phrases that occur to me throughout the day. These can arise in conversation, or in lectures, or while daydreaming. If you are thoroughly involved with the project, then they come up.

Once you have a detailed plan of what you are going to do (and you are not bound to this plan, for it will evolve as your work unfolds), then you should begin to write. Write a chapter at a time. Completely immerse yourself in each chapter. If, while writing Chapter 3, a thought occurs to you about Chapter 6, then make a note. You can, especially in the computer environment, jump

---

[1] When I was an undergraduate, one of the Assistant Professors in the Mathematics Department decided to develop his scholarly reputation by writing books. And that he did. He wrote four of them during his probation period. These were fairly advanced books, which certainly exhibited some erudition and some insight. When his tenure case came up for review, the senior faculty were *not* happy. They felt that this fellow had not fulfilled his mission. They denied the man tenure.

from one chapter to another. But the process can become confusing. Safest is to make a note—in a notebook perhaps, or in a computer file that you can pull up instantly. Then, when you begin work on Chapter 6, you have all your notes to work from.

Remember, as you write, that you are taking material that you have thoroughly digested and internalized and are presenting it to your readers—many of whom are tyros. Thus you must perform a reverse evolution to put yourself in the shoes of the learner. This may be hard to do at first, but it is a worthwhile exercise: it helps you to see as a whole how the subject is built and what questions it answers. It helps you to understand motivation and foundations.

### 5.1.3 Organization

Keep in mind that organization is a powerful tool. I have seen too many math books that state lemmas parenthetically. Here is an example:

> We thus see that every pseudo-melange is a hyper-melange. (We use here the fact that every pseudo-melange is complete. *Proof:* Let $\mathcal{M}$ be a pseudo-melange. Calculate its first Sununu cohomology group, etc.) ✠

Here the author is writing a love letter to himself. If you write such an epistle, then few will read it and fewer still will derive anything from it. You *must* present the material in its logical order. Trot out that lemma *before* you need to use it. Especially when writing with a computer, you can always add a lemma—wherever it is needed—and add suitable connecting material as well. Do not succumb to the temptation to skip this part of the writing regimen. Most of the process of developing a book consists of attending to details like making sure that all your lemmas and definitions are in place before you need them.

To recast what I have been trying to say in the last few paragraphs, the first blush of writing can be lots of fun. You organize a subject in your head, or on paper. In a flurry of enthusiasm, you write a draft on paper or on your computer. You see the subject begin to shape up as you, and only you, see it. You begin to take possession of this circle of ideas. The process is exciting and stimulating.

### 5.1.4 Hard Work

But then the moment of truth arrives. If you want to turn this random sequence of meditations into a publishable book, one that people will *read,* then some hard work lies ahead. You must go through the MS line by line, detail by detail, attending to context, syntax, logic, motivation, and many other details as well. You will proofread the same passages over and over again. Frequently, you will have to swallow your pride and rewrite an entire section, or reorganize an entire chapter. The revision process is hard, tedious work and not for the faint of heart.

You must put yourself in the shoes of the first-year graduate student, or whoever represents the ground floor of those who might read your book. Where will such a reader get hung up, and why? What can you, as the author, do to help this person along?

Finding an original way to develop the proof of the latest theorem in your subject is always a pleasure. Reorganizing that material in a new way, for your six or eight close buddies in the field, is rewarding. Much less stimulating is writing a chapter of motivation and background material. But, thinking in terms of the longevity and impact of your book, you must learn to admit that both of these tasks are of paramount importance. The latter is not going to have people buying you drinks at the next conference, but it will help your book to have an impact on the infrastructure of your subject.

### 5.1.5  Attention to Detail

To summarize, what makes for the writing of a good book is hard work and unstinting attention to detail. Frequently the work required is tedious, and you will ask yourself why you cannot assign it to a secretary or a graduate student. The answer is that you are producing *your book*, and it is for the ages, and you want it to come out right.

In some subject areas, such as literature, it is common for a professor to write only two or so books in his/her entire career. You write one to get tenure and one to get promoted to full Professor and, after that, what is the point? So each book is a major effort, and takes a number of years to complete. By the same token, each book is read *very carefully* by the author's colleagues, and taken rather seriously. It is a robust and hallowed process, and one that has stood the test of time. The writing of mathematics books is perhaps rather more seat-of-the pants. But it is just as serious. And your book can have a very distinct impact on your reputation and your career. So think carefully before launching on such a project.

## 5.2  How to Plan a Book

### 5.2.1  First Teach a Course

The business of planning a book has been touched on in the previous section. Here we flesh it out a bit.

A common way to develop a mathematics book is first to teach a course in the subject area. Indeed, teach it several times. Develop detailed notes for the course. Polish them as you go. Get your students and colleagues to read them, annotate them, criticize them. Become a good observer: note which parts of the notes make sense to your audience and which require additional explanation from you. Use these notes and observations as a take-off point for the book.

Mathematicians appear to be a shy, introspective lot.[2] It seems to exhibit

---

[2]An introverted mathematician is one who looks at his shoes when he talks to you. An

too much hubris for a mathematician to say "Now I shall write a book on thus and such." More often than not, the mathematician sneaks into the task; and a good way to do this is to develop lecture notes.

## 5.2.2 Lecture Notes

This lecture notes approach has several advantages over writing the book cold. First, you have the opportunity to classroom-test the material, to see in real time how students react to it, and to modify it according to what you learn from the experience. Second, when you teach a course you are completely involved in the material, and it is natural to develop it and revise it as you go. Third, you can show your lecture notes to colleagues—without much fear of embarrassment because, after all, they are only lecture notes—and learn from their comments and criticisms. Fourth, if the material does not seem to be developing expeditiously, you can abandon the project without losing face. After all, these were only lecture notes.

## 5.2.3 The Value of a Collaborator

It also helps to have a collaborator. Imagine going to a colleague at a conference or other group activity and saying "You know, there ought to be a book on *badeboop badebeep*." If the colleague indicates assent, then you can begin to describe what material ought to be in the book. Before long, you are swapping ideas, building each other's enthusiasm. Soon enough, you are writing a book together. Your collaborator is a reality check, and reassures you that you have not set for yourself a long-term fool's errand (for example, it would certainly be the pits to find out after two years of hard work that your book topic "Generalized Theory of Fluxions and Fluents" was no longer a matter of current interest).

Of course writing something as big as a book with a collaborator has its down side too. There will be periods when you are raring to go and he/she is busy getting a divorce, or learning to chant "Na myoho rengae kyo," or moving into a yurt. Or conversely. Taking on a book collaborator is like adding a member to your family. And the family could become dysfunctional.

## 5.2.4 Global Vision

I want to leave you with one important thought about planning a book. Try to have the entire vision of the book in place before you launch full steam into the project. Such planning enables you to keep your sense of perspective, to know how much has been accomplished, and how much remains to be done. It also helps to prevent you from wandering off onto detours, or from developing specious lines of investigation. I have written books where I have just started writing and let the course of events dictate where my thoughts would lead me.

---

extroverted mathematician is one who looks at *your* shoes when he talks to you.

Sometimes this worked well; more often it did not. After writing many books, I can say with some confidence that the planned approach is far superior.

## 5.3   The Importance of the Preface

### 5.3.1   Prefatory Remarks

I have already indicated in Section 3.4 why the Preface to any project is an important feature. For something as grandiose as a book, the Preface is paramount. Writing the Preface is part of the planning process, and it acts as your touchstone as you develop the project.

Indeed, while I am writing a book I often take a break and spend some time staring at my Preface and my Table of Contents (TOC). It may well be that, at an advanced stage of the writing, I no longer agree in detail with what the Preface and TOC say. But when I wrote the Preface and TOC my thoughts were organized and galvanized and I knew exactly what I was trying to accomplish. Studying the Preface and TOC is a way of reorienting myself.

And remember that your reviewers and your readers, if they are smart, will study your Preface and TOC in detail. These two essential front matter items will give them a preview of what they are about to read, and how to go about reading it. Just as you write the introduction to a research paper with the referee in mind, endeavoring to answer or at least minimize all his/hers observations and objections, so you write your Preface and TOC with a view that you are deflecting all the reader's *But*s.

### 5.3.2   God Is in the Details

Your Preface should not spare any detail. You have obviously thought about why existing books do not address or fill the need that your book fills. Spell this out (politely) in the Preface. You have thought about why your book has just the right level of detail and the right prerequisites. Say this in the Preface. You have thought about why your point of view is just the right one, and the points of view in other books are either outdated or misguided. Say so (diplomatically) in the Preface.

Even if you were to write your Preface, polish it to perfection, and then put it in the paper shredder, writing it would have been an important and worthwhile exercise. Writing the Preface is your (formal) way of working out exactly what you wish to accomplish with your book.

## 5.4   The Table of Contents

### 5.4.1   The Value of the TOC

In some sense, there is no way that you can know what will be in your book before you have written it. But you certainly will know the milestones, and the big ideas. In writing a novel, it may be possible to begin with "It was the best of times, it was the worst of times ..." and then let the ideas flow; however, technical writing demands more deliberation. Somehow, writing "Let $\epsilon > 0$" does not set one sailing into a disquisition on analysis. Mathematics is just too technical and too complex; you must plan ahead.

Writing the TOC is part of the early process of developing your book. It may hurt at first, and it may not feel like fun. But you will launch into writing Chapter 1 more easily if you know in advance where you are headed; conversely, if you do not know where you are headed, then how can you possibly begin? Treat the writing of the TOC like working out on your NordicTrack®: just do it.

Make the TOC as detailed as you can. The more thoroughly that you can map out each chapter and each section, the more robust your confidence will become. It will be much clearer that you can and will write this book. Always remember as you supply details that you are not wedded to this particular form of the TOC. You can, and no doubt will, change it later.

If you find yourself unable to write the TOC, then maybe God is trying to tell you something. Maybe you were not cut out to write this book or, worse, maybe you have nothing to say. Writing the TOC is an acid test. You will have to write it eventually. What makes you think that you will be able to write it *after* having written all the chapters if you cannot write it before? Does this make any sense? Write it now.

### 5.4.2 Formatting the TOC

I may note that, when you are writing in TEX, the trickiest feature is formatting. In particular, you may have trouble typesetting a Table of Contents. No worries. LATEX will do it for you. Suppose that your source code TEX file is `myfile.tex`. Simply enter the line

```
\tableofcontents
```

right after the `\begin{document}` line of your TEXfile and, when you compile, LATEX will produce `myfile.toc`. Open it up and you will see your Table of Contents.[3]

## 5.5 Technical Aspects: The Bibliography, the Index, Appendices, etc.

### 5.5.1 The Utility of LATEX

If you write your book using LATEX, or using the macros included with the book [SaK], then you have a number of powerful tools at your disposal for completing some of the dreary tasks essential to producing a good book.

---

[3]Of course this device only works *after* you have written the book. It will not help with the task that I have been describing of drafting your TOC *before* you start writing the book.

## 5.5.2   Creating the Index

In the old days, when an author created the index for a book, he/she proceeded as follows. (For effect, let me paint the whole dreary picture from soup to nuts.) First, the author sent his/her manuscript in to the publisher. For a time, he/she would hear nothing while the copy editor was working his/her voodoo on the MS. Then the publisher sent the author the copy-edited manuscript. This gave him/her the opportunity to reply to the editor's comments and suggestions. For example, the editor might have changed all the author's *thats* to *whichs* or vice versa. The copy editor might have said "You cannot call $G(x, y)$ 'the Green's function' because such usage is ungrammatical." Or "you cannot refer to 'Riemannian metrics' in Chapter 10 because, when Riemann's name came up in earlier chapters, it was not in adjectival form." (Both of these have happened to me; in the penultimate example, I was advised to call $G(x, y)$ "the function of Mr. Green.") In any event, the author slugged his/her way through the manuscript and made his/her peace with the copy editor, sometimes via a shouting match over the telephone.

At the next stage the author received "galley proofs." These were printouts of the typeset manuscript, but not broken for pages. Galley proofs were often printed on paper that was 14 inches long or more. The author was supposed to read the galleys with painstaking care, paying full attention to all details. The main purpose of this proofreading was to weed out any errors—mathematical or linguistic or formatting or some other—that were introduced by the typesetter. In particular, one would check at this stage that all the displayed mathematical formulas were set correctly.

At the next, and final, stage the author was sent "page proofs." Now the author was receiving his/her manuscript broken up into pages, and appearing more or less as it would in the final book. All figures would be in place (or at least the spaces for the figures would be in place). The pages had running heads and actual page numbers. At this propitious moment, the author was (at least in theory) no longer checking for mathematical, English, or typesetting errors. In the best of all possible worlds, at this stage a check was being made that the page breaks did not alter the sense of the text, nor did they result in figures being misplaced.

And it was at the page proof stage that the author made up the index. First, he/she went through the page proofs and wrote each word to appear in the index on a separate $3 \times 5$ card, together with the correct page reference (which was only *just now* available, since the author was working for the first time with page proofs). Then the author alphabetized all the $3 \times 5$ cards. Finally, the author typed up a draft of the index.

## 5.5.3   Using the Computer to Make the Index

In the modern, computer-driven environment for producing a book, the production process is considerably streamlined. If the manuscript is submitted to the publisher in some form of TeX, then usually the copy-edited manuscript stage and also the galley proof stage are skipped. The author works with page proofs only, and this will be his/her last "pass" over the manuscript. The entire business of writing words and page numbers on index cards, alphabetizing them, and then typing up an index script is gone. Here is the new methodology:

Imagine, for example, that you are using LATEX. You can go through your ASCII source file and tag words. (*You can do this at any stage of your writing—indeed, you may do it rather naturally "on the fly" while you are creating the book.*) For instance, suppose that somewhere in your source file the word "compact" occurs, it is the first occurrence, and you want that word to be in the index with that particular page reference. Then you put the code \index{compact} immediately adjacent (with no intervening space) to the occurrence in the text of the word "compact"; thus \index{compact} now appears in your TEX source file. (This additional TEX code does not change the printed TEX output.) There are modifications to the \index command to specify subentries in the index, and also to allow you to index items that are (strictly speaking) not words (for instance, you may wish to have \begin{document} appear in the index if you are writing a book about TEX).

You place the command \makeindex in your TEX source code file right after the \documentclass{book} command. Then, when you compile the file myfile.tex, a new file myfile.idx will be produced. This is a raw form of your index, in which the entries appear in the order in which they appear in the book—*not* alphabetized and not with subentries in place. But there will be a command in your TEX system (often called makeindex) that processes the file myfile.idx and produces yet another file myfile.ind. *That* file is the final form of your index that you can incorporate into the source code file for your book.

Just to repeat: The indexing commands cause all the words that have been marked for the index to appear in a single file, called myfile.idx (assuming that the original file was myfile.tex), together with the appropriate page references—*after* you have compiled the source file. You can then use the command makeindex to alphabetize the file myfile.idx and to remove redundancies. The procedure is documented in the LATEX book [Lam], or in the very useful book [MGBCR], or in the file makeindex.tex. (Alternatively, you can use operating system commands to alphabetize the index, and then do a little editing by hand to eliminate repetitions and redundancies. The entire process usually takes just a few hours.) The disk included with the book [SaK] also includes macros that will assist in the making of an index.

The reference [SG, pp. 76-96] treats all the technical aspects of compiling a good index. The book [Lam] has a nice discussion of the notion that you should index by *concept*, not by word. The former method allows the reader to find what he/she is looking for quickly; the latter adds—unnecessarily—to the reader's labor. A good, and thorough, index adds immeasurably to the usefulness of a book. My claim is particularly true if your book is one to which a typical reader will refer frequently and repeatedly—for example, if your book is meant to be a standard treatment of a mathematical field. Many otherwise fine mathematics books are flawed by lack of an adequate index (or, for that matter, lack of an adequate bibliography).

There are professional indexers who can produce a workable index for any book. But nobody knows your book better than you, the author. *You* should create the index. Given that modern software makes the creation of an index relatively painless, there really is no excuse for not creating one yourself.

### 5.5.4   Creating the Bibliography

Similar comments may be made about the Bibliography—this procedure has already been discussed in detail in Section 2.6. The book [SaK] tells you how to write TeX macros to compile a glossary, a table of notation, or any similar compendium. The process is rather technical, and I shall not describe it here.

### 5.5.5   Appendices

I conclude with a few words about Appendices. You will sometimes come to a point in your book where you feel that there is a calculation or a set of lemmas that you know, deep down, must be included in the book; but it will be painful to write them, and they will interrupt the flow of your ideas. The solution then is to include them in an Appendix. Just say in the text that, in order not to interrupt the train of thought, you include details in Appendix III. Then you state the result that you need and move on. This practice is smart exposition and smart mathematics as well. It is also a way of managing your own psyche: when you are attempting to tame technical material in the context of your book proper, then it becomes a burden; if instead you isolate the same material in an Appendix, then you loosen your fetters and the task becomes much easier.

An Appendix also could include background results from undergraduate mathematics, alternative approaches to certain parts of the material, or just ancillary results that are important but too technical to include in the text proper. Appendices are a simple but important writing device. Learn to use them effectively.

Many a mathematics book has an Appendix (or sometimes even a chapter) written by another mathematician. Of course that other mathematician's name appears prominently, so that he/she receives due credit. This can add a nice touch to a book, and make it appear more lively and vital.

## 5.6   How to Manage Your Time When Writing a Book

### 5.6.1   Time Is on Your Side

Many a mathematics book is started with a bang, two-thirds of it is written, the writer becomes bogged down in a struggle with a piece of the exposition, or the development of a particular theorem, and the book is never completed. I cannot tell you how often this happens; perhaps more frequently than the happy conclusion of the book sailing to fruition. I imagine that the same hangup can occur for the novelist, or for the historical writer.

I would be naive, indeed silly, to suggest that those who cannot complete their books are just insufficiently organized. Or that such people have not read and digested my advice. Anyone can develop writer's block, or can arrive at a point where the ideas being developed just do not work out, or can just lose heart. We as mathematicians, however, are accustomed to this dilemma. Most of the time, when we write a paper, things do not work out as we anticipated. The hypotheses need to be adjusted, the conclusions weakened, the definitions redeveloped. If you are going to write a book, then you will have to apply the same talents in the large. But you also need to think ahead to where the difficulties will lie and how you will deal with them. One of the advantages of doing mathematics is that nothing lies hidden. We can think and plan the entire project through, if only we choose to do so.

People in twelve-step programs, with chemical dependencies, are taught to live one day at a time. Such people are taught to concentrate on the "now." If you are writing a book, then, on the one hand, you cannot afford this sort of shortsightedness. You must plan ahead, and have the entire project clearly in view. If you kid yourself about how Chapter 8 is going to work out then, when you get to Chapter 8, you are going to pay. By analogy, if you write a paper in such a fashion that you shovel all the difficult ideas into Lemma 3, then, when it comes time to write and prove Lemma 3, you must face the music. You cannot fool Mother Nature.

## 5.6.2  Tunnel Vision

But, having said this, and having (I hope) convinced you of the value of planning, let me now put forth the advantages of tunnel vision. Once you have done the detailed planning, and you are convinced that the book is going to work, then develop an extremely narrow focus. Pick a section and write it. You need not write the sections in logical order (though there is some sense to that). But, once you have picked a section to work on, then focus on that one small task, that one small section, and do it. If some worry about another section, or another chapter, crops up, then make a note of it and then press ahead with the writing of your chosen section. Bouncing around from section to section, and chapter to chapter—chasing corrections around a never-ending vortex—is a sure path to disillusionment, depression, and ultimate failure. You can always set up scenarios for defeat. Your book-writing project can turn into a black hole, both for your time and for your psychic energy. Writing a book is a huge task; nobody will blame you if you give up, or abandon the effort. But with some careful planning, with an incremental program for progress, and with some stamina, you can make it to the end.

### 5.6.3 The Halmos Spiral Method

Paul Halmos [Ste] advocates, and describes in detail, the "spiral method" for writing a book (or a paper, for that matter). The idea is this: first you write Chapter 1, and then move on to Chapter 2. After you have written Chapter 2, you realize that Chapter 1 must be rewritten. You perform that rewrite, re-examine Chapter 2, and then you move on to Chapter 3, after which you realize that Chapters 1 and 2 must be rewritten. And so forth. If you are writing by hand, with a pen on paper, then the spiral method takes place in discrete steps as indicated. If, instead, you write with a computer, then the spiral method can take place in a more organic fashion: as you are writing Chapter 3, and realizing that Chapter 1 needs modification, you pull up Chapter 1 in another window and begin to make changes while you are thinking about them. If those changes in turn necessitate a massage of Chapter 2, then you pull it up in a third window. The advantage of doing things in discrete steps, as described by Halmos, is that you always know where you are and what you are doing; the disadvantage of the organic approach is that you can become lost in a vortex—caroming around among several chapters. The technique must be used with care.

It can only improve your work to review Chapters 1 through $(n-1)$ after you have written Chapter $n$. On the other hand, if you do use the out-of-the-box spiral method, as described and recommended by Halmos, then one upshot will be that Chapter 1 of your book will receive more attention than any other part, Chapter 2 will receive the second greatest dose of attention, and so forth (for the proof, use induction). As a result, your book *could* appear to the reader to become looser and looser as it proceeds. Perhaps this is an acceptable outcome, for only the die-hards will get to the end anyway. But when you adopt a method for its good points, also be aware of its side effects.

Certainly choose a method that works for you—organic, inorganic, spiral, or some other—and be sure to use it. If there is any time when it is appropriate to be organized, methodical, indeed compulsive, that time is when you are writing your book.

### 5.6.4 Rewriting

No matter what method you adopt for reviewing and modifying your work, keep this in mind: only wimps revise their manuscripts; great authors throw their work in the trash and rewrite. Such advice causes many to say "That is why I could never write a book; it is sufficient agony just to write a short paper." Rewriting is not so difficult; in many ways it is easier than figuring out where to insert words or to substitute passages. Treat your first try as just getting the words out, for examination and consideration. Once the thoughts are lined up in your head, then the first draft has served its purpose; you may as well discard it (and *don't peek!*). The next go is your opportunity to shape and craft the ideas so that they sing. The next round after that allows you to polish the ideas so that they are compelling and forceful. The final step allows you to buff them to a high sheen.

Use the advice of the last paragraph along with a dose of common sense. After you have struggled for a month to write down the proof of a difficult proposition, you are not going to throw it in the trash and start again. My advice here, as throughout this book, applies selectively.

### 5.6.5 Final Details

Once you have arrived at (what appears to be) the end of the task of writing your book, you still are not finished. There remains a lot of detail work. You must prepare a good bibliography (Sections 2.6, 5.5). You must prepare a good, detailed, index (the computer can help a lot here—see Section 5.5). It is generally appropriate and desirable for you to prepare a Table of Notation. You might consider building a Glossary. None of these tasks is a great deal of fun. But they will increase the value of your book immeasurably. They can make the difference between an advanced tract accessible to just a few specialists, or a book that opens up a field.

## 5.7 What to Do with the Book Once It Is Written

### 5.7.1 Putting Your Book Before the Public

You have written your *magnum opus*, slaved over it for two or more years, shown it to colleagues, received the praise of student and mentor alike. The manuscript is now polished to perfection. There is no room for improvement. Now what do you do with it?

The rules for submitting a book manuscript to a publisher are different from those for submitting a research paper to a journal. The hard and fast rule for the latter is that you can only submit a research paper to one journal at a time. Most research journals tell you up front that, by submitting a paper, you are representing that it has not been submitted elsewhere.

### 5.7.2 Submitting to a Publisher

Not so for books. You can submit a book manuscript simultaneously to several different publishers. These days there are just a few mathematics publishers— especially for advanced books. Get a feel for the different publishers by looking at their book lists. You will see what quality of books and authors they publish, and in what subject areas. Some publishers, such as, CRC Press, Springer, Birkhäuser, and the American Mathematical Society, have several book series in mathematics. Familiarize yourself with all of them so that you can make an informed choice. Talk to experienced authors to obtain the sort of information that cannot be had from advertising copy.

If you want to jump-start the publication process, then you can begin long before your book is completed. For example, if you are looking for a typing grant or an advance, then you may wish to begin negotiations with publishers after you have written just two or three chapters. Submit them, along with a Preface or Prospectus[4] (the marketing version of a Preface) and a TOC (Table of Contents). And of course include a brief cover letter saying who you are, what book you are writing, and exactly what materials you are remitting. Certainly say how long the book will be. About 400 pages is considered standard. An 800-page book may be too long. A 200-page book may be too short. It would not be out of place to also send the publisher your Vita at this time.

These days it is perfectly acceptable to send your book materials to a publisher as one or more `*.pdf` files in an email attachment. Describe in detail, in the text of your email, just what you are remitting—how many chapters, what is the subject of the book, what books it should be compared to. You can even suggest some reviewers. You should throw in a few sentences about just who you are, what your background is, and why you are the right author for this book. Attach your Curriculum Vitae to this same email.

In order to be able to negotiate intelligently with a publisher, be sure to have the following information about your book under control:

1. Subject matter and working title

2. Level (graduate, undergraduate, professional, etc.)

3. Classes in which the book could be used

4. Existing books with which your book competes

5. Working length

6. Expected date of completion

### 5.7.3  What the Publisher Needs to Know

The publisher needs to know a subject area and working title for in-house and developmental purposes. The people in the suits refer, among themselves, to the "Krantz project on fractals." So they need a working title. They need to know a working length and a sketch of the potential market so that they can price out the project. They need to know an approximate due date so that they can deal with scheduling (a non-trivial matter at a publishing house).

I am the consulting editor for a book series. One of my earliest authors completed his book two years late, with a book twice the length originally contracted; also the book was on a different subject than that specified in the contract, and with a different title. And the author wanted it to be published in

---

[4]Like a Preface, the Prospectus will describe what the book is about and why you have written it. Unlike a Preface, the Prospectus will describe the audience, the competing texts, the types of courses that could use the book, and the types of schools and departments that might adopt the book.

two volumes! I cannot tell you how much trouble I had persuading the publisher to go ahead with the project. When you are dealing with a publishing house, then you are working with business people. You must endeavor to conform to their view of the world.

## 5.7.4 Reviews of Your Book

If the publisher is interested in your project, then he/she will probably solicit reviews. Some publishers will ask you to suggest reviewers for your project. Most will not. Expect the reviewing process to take three or four months. Expect to see two to four reviews of your work.

One of the most difficult, and valuable, lessons that I have learned as an author is to read reviews. By this I mean to read them intensely and dispassionately and to learn what I can from them. Forget reacting to the criticisms. Forget justifying yourself. Forget answering the reviewers' comments. Forget melting down into an emotional puddle of goo. The point is this: even if you cannot understand what the reviewer is thinking, what he/she describes is nevertheless what he/she saw when reading the manuscript. The review describes the impression that the manuscript made on him/her. The main question you should be asking yourself as you read the reviews is "What can I learn from these reviews?" "How can I use these comments to improve my book?" There is generally something of value in even the most negative of reviews.

## 5.7.5 Evaluating the Reviews

Usually the publisher has established an initial interest in your project by looking at your Prospectus and TOC, and by agreeing to undertake the expense of reviewing (unlike a referee for a paper, a book reviewer is usually paid a modest honorarium). If the consensus of the reviews is favorable, then the publisher will most likely decide to publish your book. He/she will then ask you to take the reviews under advisement, and only that. The editor may want to discuss them with you, and may even want your detailed reaction to them. But few, if any, publishers will hold you accountable for each comment made by each reviewer.[5]

Remember this! And what I am about to say applies to research papers and to books and to anything else that you submit for review: the reviewer is not responsible for the accuracy and correctness of your work. There is only one person who bears the ultimate responsibility, and that is you. Many reviewers will do a light reading, or an overview, or will read the manuscript piecemeal, according to what interests them. If the reviewers give you a "pass," then good. But this "pass" is not a benediction, nor even a suggestion that everything you

---

[5]Note that these remarks do not apply to the writing of a textbook at the lower division level, for the so-called "College Market." Such a project is more of a team effort: you and the reviewers write the book together, in a sort of Byzantine tug-of-war procedure. The process is best learned by consenting adults in private, and I shall say nothing more about it here.

have written is correct. You must check every word, and you yourself must certify every word.

### 5.7.6   Final Revisions

In any event, the period immediately following the review process is your chance to take a couple of months and polish your manuscript yet again. (You will also have the opportunity to make small changes later on in the page proofs. But the post-review period is your last chance for substantial rewriting.) *Treat this as a gift.* It would be embarrassing to publish your book blind—with no reviews—and then to have your friends point out all your errors and omissions, or (worse) that your point of view is all wrong. The reviewing process, though not perfect, is a chance to collect some feedback without losing face and without any repercussions.

After you have polished your manuscript to your satisfaction, and presumably shown it to some friends and students and colleagues, then you submit the final, polished draft to the publisher. Many publishers will want this manuscript to be double or triple spaced, so that the various copy editors and typesetters will have room for their markings and queries. The TEX command \openup$k$ \jot, where $k$ is a positive integer, will increase the between-line spacing in your TEX output by an amount proportional to $k$.

### 5.7.7   Electronic Aspects of Publishing

Nowadays almost all of the book-publishing process is conducted electronically, and mostly over the Internet. You submit your book over email as a *.pdf file attachment. The publisher sends the *.pdf file to the reviewers as an email attachment. Each reviewer sends in his/her report as an email. The publisher removes any identifying lines from the reviews and passes them on to you (again by email). You make the appropriate edits to your TEX source file, declare the book to be finished, and send both your *.tex and *.pdf files (as well as the graphics files, if there are any) to the publisher as email attachments. It is often convenient to bundle all these files together in a *.zip file.

### 5.7.8   Electronic Sticky Notes

Then the copy editor works on your *.pdf file and marks edits on that file using "electronic sticky notes." Electronic sticky notes are a software utility that allows you to paste little notes to any page of a *.pdf file. You simply place your pointer where you want the note to be, right click on the mouse, select "electronic sticky note" from the dropdown, and you get a little yellow box in which to write your comment. The little yellow boxes become part of the file. You will read the copy editor's comments in his/her electronic sticky notes and you will respond with your own electronic sticky notes.

### 5.7.9 How the Publisher Processes Your Book

Now here is one of the great myths that exists at large in the mathematical community. People think that, in 2017, you send a `*.pdf` file to the publisher on a flash drive or an optical disc or over the Internet. The publisher puts the file in one end of a big machine and a box of books comes out the other. Technologically this phenomenon is actually possible (see Section 7.6). But a top-notch publishing house has a much more exacting procedure.

Here, instead, is what a good publishing house does with your manuscript and disc. First, an editor decides whether your book is ready to go into production. He/she may show your "final manuscript" to a member of his/her editorial board, or he/she may make the decision on his/her own. But this hurdle must be jumped. Once the book goes into production, some copy editing will be done. The actual amount will vary from publishing house to publishing house. During the copy editing process, your spelling, grammar, syntax, consistency of style, formatting, and other nonmathematical aspects of your writing will be checked. Depending on the density of corrections at this stage, you may or may not be contacted. You may have to submit another manuscript.

### 5.7.10 The Burden Is on You

One point that needs to be recorded is this. With the advent of TEX and the Internet, more of the detailed publishing burden is placed on your shoulders. When a copy editor sends you the edits for your book, it is not enough for you to say, "OK, these edits are fine by me." You have to actually go into your TEX source code file and make the edits yourself (or at least make the edits that you agree with). When you are finished, then you compile the source code file, produce a new `*.pdf` file, and send that back to the publisher.

### 5.7.11 Concerns of the Copy Editor

If you have never before written a book, then you may be surprised at the many details that a copy editor will attend to when handling your book. Here are some of these:

- All displayed equations should be formatted in the same way.

- Left and right page bottoms should align (this last task is something at which Plain TEX does not excel; LATEX handles the issue with the `\flushbottom` command).

- No page should begin with a single line that ends a paragraph (such an item is called a "widow").

- No page should end with a single line that begins a paragraph (these stragglers are called "orphans").

- Figures must be positioned properly, and rendered at the right size.

- Running heads must be checked.

- Page breaks must be checked.

- Blank pages at the ends of chapters (when the last page of text in the chapter is odd-numbered) must be completely blank.

### 5.7.12  Costs

If your project were typeset the old-fashioned way, with movable type—say that it has 400 pages—then the typesetting job would cost \$15,000–\$20,000. If instead you produce a TeX file to a level of reasonable competence, then the adjustments that I described in the last paragraph will cost \$5,000 to \$7,000. So TeX *does* save money in the publishing process.

### 5.7.13  The End of Your Role

After you have approved the page proofs, then you have reached the end of your role in the publishing process (but see the *caveat* below about the dreaded Marketing Questionnaire). The title page and copyright page and back cover copy are produced separately. I suggest that you *insist* on seeing the title page and back cover copy before the book goes to press. It happens—not often—that an author's name is misspelled or an affiliation is rendered incorrectly on the title page. Such an eventuality is embarrassing for everyone. It is best to avert it. And the back cover copy is a prominent advertisement for your book; you want to be sure that it says the right things.

### 5.7.14  Putting the Book to Bed

Even though your role is at an end, let me say a few words about what happens next. As is mentioned elsewhere in this book, when a TeX file, consisting of `ASCII` code, is compiled, then the result is a `*.dvi` file.[6] Typically, this "Device Independent File" is then translated, using software and without human intervention, to a `PostScript` file. Why `PostScript`? Many high-resolution printers read `PostScript`. Once the files for the book have been translated into `PostScript`, then the book is printed out at high resolution on RC (resin coated) paper. The result is a reproduction copy (or *repro copy*) of your book printed on glossy, nonabsorbing paper, at extremely high resolution. All the smallest subscripts and superscripts will be sharp and clear, even under magnification.

### 5.7.15  Repro Copy

The repro copy of the book is then "shot." Here, to be "shot" means to be photographed. The pages of the book are photographed onto film, in the fashion

---

[6]Although the concept of a `*.dvi` file was one of Knuth's great ideas when he created TeX, it is now the case that many systems bypass the `*.dvi` file completely and go directly to a `*.pdf` file or some other output language.

familiar to anyone who takes snapshots. But it is not printed onto photographic paper (what would be the point of that?—it is *already* on paper). Instead, the negative is then exposed or "burned" into chemically treated plates. These plates are the masters from which your book is printed. (This process is becoming ever more streamlined. Today at the AMS, the "repro copy" step is skipped altogether; the production department goes directly from the electronic file to the negative.)

Once the printing, or lithographic, plates are prepared, then the rest of the printing process—printing, cutting, and binding—is quite automated. Good books are printed sixteen pages to a sheet, and then folded and cut. This procedure results in the "signatures" that you can see in the binding of any high-quality book (not a cheap paperback). (In the old days the publisher did not cut the signatures; a serious reader owned a book knife, and did the cutting himself/herself.)

### 5.7.16 Pricing of a Book

Interestingly, the physical cost of producing a book—the printing, the binding, the cost of the paper—is well under $5 per volume; at least this is true if the book is of reasonable length (400 pages let us say) and the print run is reasonably large. The difference in cost between producing a paperback volume and a hardback volume is about $2, depending on the quality of the papers used. So why do math books cost so much?

The pricing question for books is all a matter of marketing. To be fair, the publishing house has overhead. You remember the $5,000 to $7,000 for the services of a TeXnician? This is a cost that anyone can understand. Then the salaries of the editor, the publisher, the company president, the people in the production department, the costs of marketing, the physical plant, and so forth must come out of money earned from the sale of books. Most people, indeed most authors, are not cognizant of the cost of warehousing books in a serviceable manner (so that the books are readily accessible when an order comes in). Warehousing is a fixed cost that adds noticeably to the expense of each and every book that we buy. Taxes on inventory are *very* high. These last costs are called "overhead" or "plant costs," and play much the same role as the overhead for an NSF Grant. Most publishing houses figure the cost of producing a book by taking the up-front, identifiable costs—technical typesetting, any advance to the author, print costs (often the printing is done by an outside firm), copy editing, composition, shooting—and then adding a fixed percentage (from 30% to 50%) to cover the overhead that was described above.

Then the editor does a simple arithmetic problem. He/she must make a credible, conservative estimate as to how many copies your book will sell in the first couple of years. Fifty years ago this was easy, since many academic libraries had standing orders for all the major book series. (For example, in the late 1960s, a company like Springer-Verlag or John Wiley could *depend* on library sales of 1000-1200 copies for each book!) With inflation, cutbacks, and other stringencies, libraries now pick and choose each volume. Thus the editor must make an evaluation based on **(i)** whether the book is in a hot area, like dynamical systems or wavelets or big data, **(ii)** whether people in disciplines outside mathematics (engineers, for example) will buy it, **(iii)** whether students will buy it, **(iv)** whether the author has name recognition, and **(v)** whether the book can be used in any standard classes. Other factors that figure in are **(a)** Is this the first book in an important field? **(b)** Is there stiff competition from well-established books? **(c)** How much effort is the marketing department willing to put into promoting the book? (You may suppose that the marketing department will promote any book that the editorial department sends in. On a *pro forma* level they will. But there is a delicate dynamic between these two publishing house groups, and a constant push and pull. A good editor takes pains to generate enthusiasm among the marketing people for particular books.) Having evaluated these factors, the editor writes a proposal for how many volumes of your book the house can expect to sell within a reasonable length of time (a couple of years). Then he/she figures in the company's standard profit expectation. This gives rise to the wholesale price of the book.

As an example, suppose that you write a book on a fairly specialized area of partial differential equations. After an analysis of the foregoing kind, the editor determines that the book is sure to sell 500 copies in the first two years. The up-front costs are \$15,000. Add 50% for overhead and that makes \$22,500. Add 20% for the company's standard profit margin and that brings the total to \$27,000. The wholesale price of the book must, after sales of 500 copies, bring in that much money. (If a given editor has several books that fail to meet this simple criterion, then he/she is out of a job.) Now do the arithmetic. You will find that the wholesale price of this book must be \$54 per volume. Thus a bookstore will probably sell it for at least \$70 to \$80. Now do you understand why mathematics books cost what they do?

Incidentally, if the difference in cost between producing a hardcover copy of a given book and a paperback copy of that same book is about \$2, then what accounts for the large difference in price between hardcover and paperback books? The answer, apart from marketing voodoo, is that the costs of producing the book tend to be covered by the sale of the hardcover version. Thus the publisher has considerable latitude in pricing the paperback edition. John Grisham novels stay in hardcover format for more than one year before the paperback edition is released; usually, the hardcover edition sells millions of copies. The production costs, and the huge advance that Grisham garners for each of his books, are well covered by the hardcover sales. Thus the publisher is ready to make real money when the paperback edition is released. He/she can be imaginative both in pricing and in production values—if the physical cost of producing a volume is \$5, then he/she can price it for as little as \$7-\$15 and expect to sell a great many copies.[7]

---

[7]Interestingly, the entire notion of mass market paperbacks was invented in the early 1950s by Mickey Spillane and his publishers—Dutton and Signet. By 1955, Spillane had written three of the five best-selling books in history—and he had only written three books! By contrast, Margaret Mitchell's blockbuster *Gone with the Wind* [Mit] sold fewer than a million copies in its first two years—all in hardcover, of course. James Gleick's *Chaos* [Gle] has sold

### 5.7.17 Marketing Your Book

Back to math books. In the preceding discussion there was an important omission. How does the editor make the market determination that I described three paragraphs ago? He/she can always consult his/her editorial board and his/her trusted advisors. But let me reassure you that he/she will certainly study your Prospectus and Preface, and he/she will pay close attention to your *Marketing Questionnaire*.

The latter item bears some discussion. Whenever you write a book for a commercial publishing house, and often for a professional society or a university press, you will be sent a Marketing Questionnaire to complete. I hate to complete these things, and you will too. But you must do it. I have heard authors say "I'll just phone the editor and talk to him about this stuff." Sorry; that just will not do. You must complete the questionnaire, and carefully.

What is this mysterious object? First, the questionnaire is long—often 10 pages or more. Second, it asks a lot of embarrassing questions: What is your hometown newspaper? Which professional societies might be interested in your book? What are the ten strongest features of your book? What is the competition? Why is your book better? In which classes can your book be used? What is typical enrollment in those classes? How often are they taught?

As mathematicians, we are simply not comfortable fielding questions such as these. We do not think in these terms. But, if you have been attending to the message of this section, then you can see how an editor can use this information to help price out the book. So why can you not just go over this stuff on the phone with the editor? One reason is that the editor needs this information *in writing*—for the record, and to show that he/she is working from information that *you* provided, and for future reference. The other is that the questionnaire will be passed along to the marketing department for the development of advertising copy and marketing strategies for your book. Like it or not, the Marketing Questionnaire is important. Take an hour or so and fill it out carefully.

### 5.7.18 Author Frustrations

When I was developing my first book, and negotiating with my publisher, I asked the editor what I would be peeved about three years down the line. He told me that I would be unhappy about the size of the print run, and I would be unhappy with the advertising. Then he explained to me how the world works. First, think about the sales figures that I described above. And think about the fact that a business must pay a substantial inventory tax for stock on hand. Extra books sitting around are a liability. And today (with new printing technology) small print runs are not so terribly expensive as they were even ten years ago. Even print-on-demand is feasible in many cases (see Section 7.6). So if the publisher thinks that your book will sell 500 copies in the first couple of years, then the first print run is likely to be only 750. When that stock starts to

---

about the same.

run low, another 750 can be generated easily. The money saved per unit with a print run of 1500 (as opposed to 750) is relatively small, and is sharply offset by storage costs and inventory tax.

### 5.7.19 Advertising Your Book

And now a word about advertising. There is nothing that an author likes better than to open the *Notices of the American Mathematical Society* or the *Mathematical Intelligencer* or the *American Mathematical Monthly* and to see an ad for his/her book. Of course a full page ad is best (and almost never seen), but a half page, or quarter page, or even an ad shared with eleven other books, is just great. Typically, you will see such an ad just once for your book. After that, your name and the title of your book will appear in the company's catalogue. Of course there will be advertising material online. For a textbook there could be an entire website devoted entirely to one book. Many publishers rely on "card decks"—stacks of $3'' \times 5''$ cards, each with a plug for a single book—that are mailed in a block to mathematicians. Usually the potential buyer can mail in a card, without money, and receive a copy of a particular book for a 30-day examination period.

### 5.7.20 Book Contracts

In the spirit of doing first things last, let me now say a few words about book contracts. When a publishing house is interested in publishing your book, then it will send you a contract. Typically, you will be offered a royalty rate of 10% to 15%. You will be given a submission deadline, and this deadline is definitely negotiable. Err on the conservative side (more time, rather than less), so that you have a fighting chance of finishing the book on time. If you do finish on schedule, then the publisher will take a shine to your project. If you do not, and the project is six months late, then most publishers will be forgiving; but, technically, a late project is no longer under contract!

A rough page length will be specified in the contract, and a working title given. Sometimes you will be offered an advance against royalties, or a typing grant. Sometimes you will be asked to certify that you will submit your manuscript in some form of TEX. Then there will be a lot of legal gobbledy-gook, most of which seems to be slanted in favor of the publisher. For the most part, it is. The publisher wants to be able to pull the plug on a project whenever and wherever it deems such an action suitable. Honorable publishers do not like to exercise this option, but they want to have the option available.

I can tell you that many authors—especially first-time authors—are quite uncomfortable with standard book contracts. This uneasiness stems, for the most part, from lack of familiarity. The details of the contract *can* be negotiated, and you should discuss with your editor any passages or provisions that you do not like. If the royalty seems too small, then negotiate. If the publisher wants *you* to render the artwork in final form, and you cannot or will not do it, then negotiate. If you do not like the deadline, then negotiate. If the number of gratis copies of the work offered to the author is not adequate, then negotiate some more. Usually such negotiations are fairly pleasant. You will find the editor eager to cooperate—as long as your demands are within reason.

### 5.7.21 The *Textbook and Academic Authors Association*

You may find it attractive to join the *Textbook and Academic Authors Association* (TAA).[8] This organization was formed to defend the rights of authors, and to teach authors about the publishing process, and will help you in dealing with publishers. It also has a rather informative newsletter. And membership gives you access to a number of useful discounts, so that your dues are almost a wash.

I have dealt with many publishers. Most of them are very good to their authors (as well they should be) and most employ knowledgeable and competent editors. However, forewarned is forearmed. It is helpful to be familiar with the publication process before you launch into it.

## 5.8 Royalties

### 5.8.1 Compensation for Your Work

It makes sense that the author of a book will want to be compensated for his/her efforts. In other words, the author expects some royalties. Of course a math book is not going to sell like a Tom Clancy novel. But one can make a nontrivial amount of money from a math book. As an instance, calculus author Jim Stewart built a $26 million dollar home in Toronto with his royalties.

These days the royalty rate for an undergraduate text ranges from 10% to 15%. It could be considerably more for a well-established author with a popular book. For a graduate text or monograph, the royalty could be less. Here is the passage from a recent contract for an upper-division math text going into its fourth edition (this is in fact a real analysis text):

---

6. ROYALTIES

(a) The Publisher agrees to pay the Author (or someone designated by the Author), and the Author shall accept as payment in full for writing and delivering the Manuscript, Illustrations, and index, for the performance of all of the obligations of the Author hereunder, and for all the rights granted to the Publisher pursuant to this Agreement, the following amounts:

  (i) For copies in print or eBook format sold by the Publisher in the United States of America, twelve percent (12%) on the first 750 copies

---

[8]*Textbook and Academic Authors Association*, P. O. Box 367, Fountain City, Wisconsin 54629. The web address is `www.taaonline.net`.

and fifteen percent (15%) thereafter of the Publisher's net receipts (as defined in Paragraph **6(d)** below).

(ii) On translations, licensing sales, electronic database sales, excerpts, abridgments, deep discount sales (sales at a discount of fifty percent (50%) or greater of the Publisher's established list price of the Work), and on all sales of copies of the Work outside the United States of America, the Publisher shall pay royalties at one-half (1/2) the rate set forth in Paragraph **6(a)(i)** above in respect of the Publisher's net receipts. In the event the Work is included in an electronic database with other works, or is otherwise exploited in combination with other works, royalties shall be apportioned by Publisher in its sole discretion, exercised in good faith.

**(b)** In the event the Publisher exercises any of the rights of the Publisher pursuant to Paragraph **5** above and a royalty is not specifically provided for, the royalty which shall be payable to the Author shall be one-half (1/2) of the rate set forth in Paragraph **6(a)(i)** above in respect of the Publisher's net receipts.

**(c)** Notwithstanding the above, no royalty will be paid on copies of the Work furnished gratis for review, advertising, promotion, bonus, sample, or like purposes, or on copies of the Work sold at less than Publisher's cost, or on any copies returned to Publisher for any reason, or on copies of the Work sold to the Author. Free use of the rights granted herein may be made by the Publisher to promote the sale of copies of the Work and the rights therein. The Publisher may create a reasonable reserve for returns when calculating royalties.

**(d)** For purposes of this Agreement, the Publisher's "net receipts" from sales shall mean monies received by the Publisher from such sales less adjustments for discounts, credits, and returns. Royalties will not be paid on prepaid transportation, postage, insurance, and taxes. The Publisher's "net receipts" from licensing or assignment shall mean monies received by the Publisher less any specified costs of such licensing or assignment.

**(e)** All payments made under the terms of this Agreement will be subject to Federal income tax withholding, as required by the United States Internal Revenue Code.

**(f)** All royalties and other income accruing to the Author under this Agreement shall be credited to an account maintained on the records of the Publisher (the "Royalty Account"), which Royalty Account will be charged for all amounts paid or payable to Author, including any advance payments, and for all amounts Author is charged, or obligated to pay, pursuant to this Agreement.

You can see that the publisher is careful to cover all possible scenarios, and that the contract is written so that no misunderstanding is possible. The publisher is also very explicit about royalty rates for *e*-books, for electronic databases, and other high-tech versions of the book.

## 5.8.2 Negotiating the Royalty Rate

It is possible to negotiate the royalty rate with the publisher. I once retained a publishing attorney to negotiate publishing contracts for me. He got me some terrific royalty rates, but afterwards the publishers were quite annoyed with me for having indulged in this artifice. It seems that it is OK for the editor (in the course of the negotiations) to say "I'll have to check with our lawyers," but it is *not* OK for me (the author) to say it.

# Chapter 6

# Writing with a Computer

*Computers are useless. They can only give you answers.*

Pablo Picasso

*If he wrote it he could get rid of it. He had gotten rid of many things by writing them.*

Ernest Hemingway
*Winner take Nothing* [1933]. Fathers and Sons

*Easy reading is damned hard writing.*

Nathaniel Hawthorne

*In a very real sense, the writer writes in order to teach himself, to understand himself, to satisfy himself; the publishing of his ideas, though it brings gratification, is a curious anticlimax.*

Alfred Kazin

*On seeing a new piece of technology:*

*A science major says "Why does it work?"*
*An engineering major says "How does it work?"*
*An accounting major says "How much does it cost?"*
*A liberal arts major says "Do you want fries with that?"*

Anon.

*[With reference to Germany] One could almost believe that in this people there is a peculiar sense of life as a mathematical problem which is known to have no solution.*

Isak Dinesen

## 6.1   Writing on a Computer

### 6.1.1   Technology and Writing

Today most every mathematician writes on a computer. You may find it cathartic to generate your early drafts in the old-fashioned way— by writing by hand with a pen. Fine. Many of the best writers create their work in that fashion. But, in the end, your paper or book will be rendered on a computer. This is the world we live in.[1]

Clearly, when you are writing on a piece of paper with a pen or pencil, then you can easily and naturally jump from one part of the page to another. You can, in a comfortable and intuitive fashion, jot marginal notes and make insertions. You can put diacritical marks and editorial marks where appropriate. You can scan the current page, flip ahead or back to other pages, sit under a tree with your entire MS clutched in your fist, put `Post-It` notes in propitious locations, tape addenda to pages, and so forth.

Now the fact is that almost all the "old-fashioned" devices described in the preceding paragraph have analogues in the computer setting. And the computer has capabilities that the traditional milieu lacks: magnificent search facilities, unbeatable cut-and-paste features, the power to open several different windows that either contain several different documents or several different parts of the same document, and many others as well. With a computer, you can have your text open in one window, a dictionary open in another window, and the Internet open in a third window. What could be better?

But you still must use the tools that work for you. If you have been writing with a pen on paper for many years, then you may be disinclined to change. At a prominent university on the east coast, there was an eminent and prolific mathematician, who had access to any writing facilities that one might wish, and who wrote by candlelight with a quill. That was his choice, and it certainly worked for him.[2]

In this section I want to say a few words about writing on the computer, and what I find advantageous about it. I am addicted to writing on the computer. It makes me more productive and efficient, and it saves me a lot of time and aggravation. You must decide for yourself whether you want to write on the computer, and whether the computer is the system for you. Famous and successful novelist John Irving is dyslexic. He writes only by hand, and hires a typist to type up his manuscripts. This practice works for him. You have to decide what works for you.

I will reserve comments about specific writing systems, like TeX and `Word`, for a later section.

## 6.1.2   Writing on a Computer

When writing on a computer, you can type as fast as you wish, never fearing for spelling or other errors (just because making corrections is trivial). When you become acclimated to the medium, you can create text at least as fast as you would have with a pen (assuming that you know how to type), and the text

---

[1] I do have two friends who write *everything by hand*. When the handwritten draft is complete, they pay a TeXnical typist about $8 per page to render the project as a TeX document. One of these people even submits book projects to publishers in handwritten form. In today's world, these two people can be considered exceptional if not admirable.

[2] He also ate dinner every night wearing a tuxedo.

will always be legible. You can make corrections, insertions, and deletions, and move blocks of text with blissful ease. You can print out beautiful paper copy of your work (paper copy is called *hard copy*), and you can store your work on your hard disk or hard drive (also known as the *fixed disc*).[3] You never need worry about misplacing all or part of your manuscript, since finding files on your hard drive is trivial.[4]

### 6.1.3 Finding Files on Your Hard Drive

It is particularly easy to find things on your hard drive. Even if you have forgotten the name of the file, or the size of the file, or what the file is about, you use either `Windows` functions or utilities easily obtained online to find your stuff. If the only thing you can remember about your document is a word or phrase in it, you will be able to find the file containing the document in seconds.

To illustrate this last point, I often find myself printing out another copy of a paper or chapter that I am working on, rather than trying to find where I put my last paper copy. I can find my file on my hard disk and print it out in just a moment; the old approach, more traditional and agonizing, of searching through my study for my hard copy could take hours. And remember this point: any tool that prevents your writing moods from being interrupted or jarred is a valuable one. My computer has eliminated, for me, the need to search my office for the paper copy that I want to work on. It saves me hours of time, and it saves me considerable irritation. Cherish those tools that make your life easier, and learn to use them well.

### 6.1.4 Multiple Versions of Your Work

When working on a computer, you can easily keep every single version of a document you are writing. Suppose, for example, that you are writing an

---

[3]Today many computer systems have multiple external drives (hard, flash, and other). I store all my work on external drives—never on the fixed disc.

[4]Noted computer expert Peter Norton got his start by rescuing Hollywood writers who would lose their work on their computers. Most people do not realize that, at least on a `Windows` machine, when you erase a file then it is not really erased. In fact the first character of the name is changed to an exclamation point.

In more detail, `Windows` (and other operating systems) keeps track of where files are on a hard drive through pointers. Each file and folder on your hard disk has pointers that tell `Windows` where the file's data begins and ends. When you delete a file, `Windows` removes the pointers and marks the sectors containing the file's data as available. From the file system's point of view, the file is no longer present on your hard drive and the sectors containing its data are considered free space. However, until `Windows` actually writes new data over the sectors containing the contents of the file, the file is still recoverable. The reason that the operating system does things this way is that it is quick. Actually *erasing* a file could take several minutes. Removing the pointers just takes a second.

A file recovery program can scan a hard drive for deleted files and restore them. This was Peter Norton's innovation in the early 1980s. The main point is that the file is still there, even though you think you have erased it. Norton understood this point, and wrote proprietary software that enabled him easily to identify and recover erased files. This led to his creation of the noted software `Norton Utilities`.

article about diet fads among troglodytes. The first version of your article could be called `TROG.09-21-17.001`. After you modify it, the second version could be called `TROG.09-22-17.002`. The third would then be called `TROG.09-24-17.003`. And so forth. All these would be neatly stored, and accessible, on your hard disc. Notice that I build the date of creation into my name for each file. This is because I do not want to be at the mercy of the automatic dating system built into the computer operating system.

Compare with the situation, in a paper office, in which you had several different versions of a document. How would you store them all? How would you keep track of and differentiate among them? How would you access them? Note that a computer also assigns a time and date stamp to each file you process. Thus, when you do a directory reading, you would see something like this:

```
TROG.09-21-17  001        2357     09-21-17       11:15pm
TROG.09-22-17  002        3309     09-22-17        2:31pm
TROG.09-24-17  003        3944     09-24-17       10:42pm
TROG.09-29-17  004        4511     09-29-17        9:11am
TROG.10-02-17  005        3173     10-02-17        2:04am
```

We see here five versions of the paper. The third column shows the number of bytes in each version. The fourth shows the date on which the editing of that version was completed. The last column shows the exact time of completion.

Note that the versions grew in size until, in the wee hours of October 2, the author decided to discard more than 1338 bytes of the document; this resulted in version 005. Is it not reassuring to know that all the old versions are available, just in case the author decides to resuscitate some of his/her old turns of phrase? Whether or not you are in the habit of examining old drafts of your work, you will find it psychologically helpful to have all the old versions. When the work is complete you can, if you wish, discard all the drafts but the final one. But the fact is that mass storage space is so cheap and plentiful these days that every draft of every one of your works, even if you are Stephen King and Tom Clancy rolled into one, will only take up a small fraction of your hard disc.

### 6.1.5   Text Editors

If you do your writing with a first-rate text editor, as I do, then you have powerful tools at your disposal (see [SaK] and also our Section 6.3 for a discussion of text editors). You can open several files simultaneously, have several different portions of the same file open at the same time, and have a bibliographic resource file open; with an environment like `Windows`, you also can have an optical drive or online thesaurus and dictionary open and also be connected to `Google` on the Internet—and you can jump from one setting to the other effortlessly. Given that any trip to the dictionary could take five or ten minutes the old-fashioned way, and ten seconds the electronic way, think of how much time you will save over a period of several years. Again—and here is *the* most important point—by using technology you circumvent the danger of your thought processes and your creative juices being interrupted.

Even though I am addicted to writing with a computer, hard copy plays an important role in my writing process. For, after I have written a draft, I print it out, lounge in my most comfortable chair, and proofread and edit. There exist methods of proofreading and editing directly on the computer—I shall not go into them here. But, because of my age and my training, I find that there is nothing like a paper copy and a red pen to stimulate critical thinking. You will have to decide for yourself what works for you.

### 6.1.6 Handwritten Manuscripts

There is a down side to writing with a computer; you can work your way past this one, but you had best know about it in advance. When you create a document on a computer system—especially if you use a sophisticated computer typesetting system like TeX (see Section 6.5)—then the printed copy looks like a finished product. This makes it even more difficult than usual for you, the author, to see the flaws that are present. Even with handwritten copy you will have difficulty seeing that certain paragraphs must go and others must be rearranged or rewritten. But, when the MS is typeset, the product looks etched in stone. One cannot imagine how it could be any more perfect. Believe me, it can always stand improvement. You will have to retrain yourself to read your typeset work critically.

(For the flip side of the last paragraph, consider this. I was recently asked, by an important publishing house, to evaluate a manuscript for a textbook that they were considering developing. The manuscript was *handwritten.* This flies in the face of all that is holy; a manuscript going to a publisher should always be typed or word processed. In any event, I took what I was given and wrote my report. But this was a difficult process for me. I had to keep telling myself that this was *not* a rough set of notes, that it was a polished manuscript—even though it was handwritten and *looked* like a rough set of notes. Play this paragraph off against the last one for a lesson about form over substance.)

### 6.1.7 Backups

And now a coda on backups. If you use a computer for your work, then develop the habit of doing regular backups. The "by the book" method for doing backups is generally agreed to be the "modified Tower of Hanoi" protocol. Here the Tower of Hanoi is an ancient game as shown in Figure 6.1. The goal is to move the stack of wooden discs—one disc at a time—from the first spindle to the last spindle *without ever placing a larger disc atop a smaller disc.* The logic of the Tower of Hanoi backup protocol is based on the logic of the game.

The Tower of Hanoi rotation method is one of the more complex backup schemes. It is based on the mathematics of the Tower of Hanoi puzzle, using a recursive method to optimize the back-up cycle. Imagine that you are backing up to tapes. Every tape corresponds to a disk in the puzzle, and every disk movement to a different peg corresponds to a backup on that tape. So the first

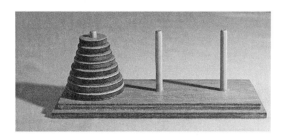

Figure 6.1: The Tower of Hanoi|Plywood*.

tape is used every other day $(1, 3, 5, 7, 9, \ldots)$, the second tape is used every fourth day $(2, 6, 10, \ldots)$, the third tape is used every eighth day $(4, 12, 20, \ldots)$.

A set of $n$ tapes (or other media) will allow backups for $2^{n-1}$ days before the last set is recycled. So, three tapes will give four days' worth of backups and, on the fifth day, set $C$ will be overwritten; four tapes will give eight days, and set $D$ is overwritten on the ninth day; five tapes will give 16 days, etc. Refer to Table 6.2. Files can be restored from 1, 2, 4, 8, 16, ..., $2^{n-1}$ days ago. This gives you access to any configuration that your hard disk has had for the past several weeks.

**Days of the Cycle**

|  | 01 | 02 | 03 | 04 | 05 | 06 | 07 | 08 |
|---|---|---|---|---|---|---|---|---|
|  | A |  | A |  | A |  | A |  |
| **Set** |  | B |  |  |  | B |  |  |
|  |  |  |  | C |  |  |  | C |

Table 6.2. The three-tape Tower of Hanoi Schedule

Not all of us are up to this level of rigor. But do *something*. At least once per week, back up all your critical files to an external hard drive, a flash drive, a DVD, a Blu-Ray disc, or some other mass storage device. Losing your C: drive is analogous to having your house burn down. It is an experience that you can well do without. Regular backups on different media are a nearly perfect insurance against such a calamity.

## 6.1.8 Using the Cloud

These days it is a good yoga to back up your files to the Cloud. On the one hand, the Cloud is merely somebody else's hard drive. No different from your own hard drive. On the other hand, the Cloud is very well maintained and is accessible from anywhere—with a notebook computer, a tablet, or a smart cell

---

*The photograph of the Tower of Hanoi in plywood was taken from commons.wikimedia.org/wiki/File:Tower_of_Hanoi.jpg. Permission to use this image granted under the GNU Free Documentation License.

phone. And it is extra security for your data. Many hard drive manufacturers provide free access to Cloud space. Your home institution may offer you Cloud space (mine does—I have 50 Gigabytes of Cloud at Washington University). Or, if necessary, you can buy Cloud space for a nominal fee.

In fact, Mac OS X users are now offered a service that will automatically back up files in the `Documents` folder to Apple's Cloud server. Also, the Cloud copy is automatically updated whenever a change is made to a file in the `Documents` folder. Both Apple and Microsoft support a once-an-hour or once-a-day backup of all files to one's external disk. One can hardly ask for a better insurance policy against data loss. In addition, these free services are well organized and easily accessible.

## 6.2 Word Processors

### 6.2.1 Processing Words

I have already indicated in Section 6.1 the advantage that working on a computer has to offer. Next I shall specialize down to word processors and what they do.[5] I do this in part so that, when you read Section 6.5 about TEX, you will appreciate the differences.

A word processor is a piece of software—first developed by Charles Simonyi and a group of software developers in 1974.[6] You use this device for entering text on the computer screen, and for saving the text on a storage device (usually a disc). You engage in this process by striking keys on a keyboard—very similar to typing. The word processor performs many useful functions for you:

1. When you get to the end of a line, the word processor jumps to a new line—you do not have to listen for a bell, or keep one eye on the text, as you did in the days of typewriters.

2. The word processor allows you to insert or delete text, or to move blocks of text from one part of the document to another, with ease and convenience. You can create a new document (such as a letter) by making a few changes to an existing document.

3. The word processor right justifies (evens up the right margin) of your document. This process results in a more polished look.

4. The word processor can check your spelling.

5. The word processor communicates with your printer, and ensures that the document is printed out just as it appears on the screen (this is what we call `WYSIWYG`, or "What you see is what you get.").

---

[5]From the point of view of a modern mathematician, word processors are really old school. No serious mathematical writer uses a word processor. There are so many other truly excellent alternatives available. But I take a little time to describe the features of word processors to help the reader to put things into context.

[6]In fact their product was called `Bravo`.

6. The word processor enables you, if you wish, to incorporate graphics into your document.

7. The word processor allows you to perform "global search and replace" functions. For example, if you are writing a paper about mappings, and you decide to be hypermodern and start calling them functors instead of mappings, then this change can be made *throughout the paper* with a few keystrokes.

8. The word processor allows you to select from among several different fonts: Roman, boldface, italic, Dunhill, Lucida, Times New Roman, typewriter-like, and so forth.

## 6.2.2  Shortcomings of Word Processors

These days, most professional people and most business people prepare their documents on a word processor. Using a word processor saves time, money, and manpower. From the point of view of a mathematician, a word processor is not entirely satisfactory. The primary reason is that a word processor will not typeset mathematics in an acceptable fashion. A typical word processor can display *some* mathematics, but not in a form similar to what you would see in a high-quality book. The word processor cannot treat complicated mathematical expressions: a commutative diagram, the quotient of a matrix by an integral, or a matrix with entries that are themselves matrices. It cannot render characters in all the different sizes needed. It cannot kern.

Even for simple mathematical expressions, such as a character with both a superscript and a subscript, the output from a word processor is nowhere near the quality that one would see in a typeset book. Microsoft $\text{Word}^{\circledR}$ is today the most prominent word processor, and it does have notable mathematical capabilities. There are serious scholars who use Microsoft Word in preference to TeX. But they are definitely in the minority.

There are patches you can buy—for Microsoft $\text{Word}^{\circledR}$, for instance—that enable some mathematical formulas. Among these are MathType and MathML. But they are of nowhere near the quality that TeX outputs. Certainly not the quality of a finished book. When it comes to delicate matters of kerning and other spacing and formatting issues, word processors are limited in their abilities. And this is in the nature of things, just because a word processor is WYSIWYG and uses monospaced fonts.

Outside of mathematics—in the *text*—word processors fall short in that they do not *kern* the letters in words; many word processors use monospaced fonts, just like a typewriter. This fact means that the word processor does not perform the delicate spacing between letters—spacing that *depends* on which two letters are adjacent—which is standard in the typesetting process. The word processor does not offer the variety of fonts, in the necessary range of sizes, ordinarily used in typesetting. The word processor does not have sufficient power to make fine adjustments to horizontal and vertical spacing on the page—both essential for the demands of quality page composition.

Put in other terms, a word processor is constrained by the fact that it is `WYSIWYG` ("what you see is what you get"). Even a high-quality screen is no more than 400 pixels (dots) per inch, while high-quality printing is 2400 dots or more per inch. Since a word processor prints *exactly* what appears on the screen, it can format with no more precision than what can be displayed on the screen. TEX, by contrast, is a markup language. It gives typesetting and formatting *commands*. It can position each character on the page within an accuracy of $10^{-6}$ inches.

### 6.2.3   A Word Processor is `WYSIWYG`

Because a word processor is `WYSIWYG`, any file produced by a word processor will contain hidden formatting commands.[7] One side effect of this simple fact is that if you cut out a piece of text from a word processor file and move it to another part of the file, then it may not format properly. As an instance, suppose that you have a displayed quotation (such a display usually has text with wider margins and space above and below). An example is

> Four score and seven years ago our fathers brought forth on this continent, a new nation, conceived in Liberty, and dedicated to the proposition that all men are created equal.
>
> Now we are engaged in a great civil war, testing whether that nation, or any nation so conceived and so dedicated, can long endure. We are met on a great battle-field of that war. We have come to dedicate a portion of that field, as a final resting place for those who here gave their lives that that nation might live. It is altogether fitting and proper that we should do this.

Snip that out using standard commands for your word processor or operating system and drop it in elsewhere in your document; it will not format correctly and you will waste a lot of time fixing it up. Because TEX is a markup language, it does not suffer this formatting malady. One of the beautiful features of TEX is that you can cut and move a fantastically complicated display and it will not change one iota.

### 6.2.4   Non-Universality of Word Processors

Finally, no word processor is universal. There are too many word processing systems. They are all compatible to a degree, but not in the way that they treat

---

[7]You may find the following experiment to be interesting. Use a word processor to create a file, called `hello.doc`, containing just the sentence "Hello world." Next use a text editor to create a file, called `hello.txt`, that contains only the sentence "Hello world." You will note that the indicated sentence has ten letters, one period, and one blank space. So you would guess that a file containing the sentence would have 12 or so bytes. For the text editor it is true. For the word processor definitely not. In fact such a file created with Microsoft `Word` has nearly 10,000 bytes. Why? It is the hidden formatting commands!

mathematics. Thus, again, if one is doing mathematics using a word processor, then one will be hindered. Your ability to communicate with, and to collaborate with, other mathematicians, will be limited.

## 6.3   Using a Text Editor

### 6.3.1   What Is a Text Editor?

Text editors are, primarily, for the use of programmers. A programmer wants an environment for entering computer code; the code will later be *compiled* by a JAVA compiler, a PHP compiler, a C++ compiler, or some other compiler. Thus a text editor should not perform value-added features to the code that has been entered: there should be no hidden (binary) formatting commands, no hidden (binary) instructions for the printer, no hidden bytes, or any other secondary data. A file created with a text editor should comprise only the original `ASCII` code, together with space and line break commands.

### 6.3.2   Other Tools

In today's world there are rather more sophisticated tools, like the open source environment `Eclipse`. In the words of `Eclipse`

> `Eclipse` provides integrated development environments (IDEs) and platforms for nearly every language and architecture. We are famous for our Java IDE, C/C++, JavaScript and PHP IDEs built on extensible platforms for creating desktop, web and Cloud IDEs. These platforms deliver the most extensive collection of add-on tools available for software developers.

For the purposes of doing TEX, a standard text editor is sufficient for most people. The text editor that comes with PC-TEX is customized for TEX users. And a text editor like `Crisp` can be customized by the user for TEX or for any other application.

The text editor `WinEdt` is a self-contained environment for creating LATEX documents. This includes handling the graphics. Here is what `WinEdt` says about itself:

> Although reasonably suitable as an all-purpose text editor, `WinEdt` has been specifically designed and configured to integrate seamlessly with a TEX System (such as MikTEX or TEX Live).[8]   However, `WinEdt`'s documentation does not cover TEX-related topics in depth; you'll find introductions and manuals on typesetting with TEX, as

---

[8]TEX Live is a free software distribution for the TEX typesetting system that includes major TEX-related programs, macro packages, and fonts. It is the replacement of its no-longer-supported counterpart teTEX. It is now the default TEX distribution for several `Linux` distributions such as `openSUSE`, `Fedora`, `Debian`, `Ubuntu`, and `Gentoo`. Other `Unix` operating systems like `OpenBSD`, `FreeBSD`, and `NetBSD` have also converted from teTEX to TEX Live.

well as links to other recommended accessories, on TEX's Community Site (TUG). For LATEX-related issues visit LATEX Community Forum: questions are welcome and help is forthcoming!

These types of tools are constantly evolving. In ten years `Eclipse` could easily be surpassed by some newer technology.

### 6.3.3 Files Created on Text Editors

A document printed directly from a file created with a text editor would look just like what you see on your computer screen—typewriter-like font and all—with a ragged right edge and with old-fashioned monospacing. Such a document might be acceptable for an in-house memo, but it is not formatted in a manner that would be suitable for public use. Thus why would a mathematician want to use a text editor?

### 6.3.4 The Importance of TEX

Today, TEX is the document creation utility of choice for mathematicians (see Section 6.5). Apart from its flexibility and the extremely high quality of its output, TEX is also infinitely portable, and it is the one system that you can depend on most (and soon all) mathematicians knowing. If you want to work with a mathematician in Germany, using the Internet, and if you were to say to your collaborator "let's use `OpenOffice`®," then you would be laughed right off the stage. The only choice is TEX (or one of its variants, such as LATEX). And the point is this: TEX is a high-level computing language (and also a *markup* language—see Section 6.5). You create a TEX document using a text editor..

Many a TEX system comes bundled with its own text editor. Usually such a bundled editor has many useful features that make it particularly easy to create TEX documents. If you are a PC user, however, then you are accustomed to selecting your own software. `Windows` comes with a perfectly serviceable text editor called `NotePad`; many popular word processors, such as `OpenOffice` and `Word`, have a "text editor" mode. But much more sophisticated text editors may be purchased commercially. One of the best is `Crisp`® (which is a version of the `UNIX` text editor `emacs` that has been adapted for the PC). A good text editor can be customized for specific applications, allows you to open several documents and several windows at once, has sophisticated search and cut-and-paste operations, and will serve you as a useful utility.

### 6.3.5 Variants of the TEX Software

TEXShop is a free LATEX and TEX editor and previewer for `OS X`. It is licensed under the `GNU GPL`. TEXShop was developed by the American mathematician Richard Koch. TEXShop was modeled on `NeXTstep`'s bundled `TEXview.app` and developed for the then new `OS X` user interface `Aqua` and capitalized on the native `PDF` support of that version of the Apple operating system; this was itself based on `NeXTstep`'s successor `OpenStep`. Mitsuhiro Shishikura enhanced it by adding the ability to transfer mathematical expressions directly into `Keynote` presentations. Lacking the TeX `eq -> eps` Service which TeXview afforded, other apps such as LaTeXiT.app were developed to provide Service support. TeXShop requires an existing TeX installation and is currently bundled with the MacTeX distribution.

TeXShop (then version 1.19) won the 2002 Apple Design Award of Best Mac Open Source Port for its capability to display scientific and technical documents created in TeX format. In fact, TeXShop makes it possible, thanks first to `pdfsync.sty`, to switch back and forth between code and preview easily, jumping at a corresponding spot, simply by a CMD-click. From TeXShop 1.35 onward this also works with multipart documents, which are joined by `\include`. Also, with version 1.35, TeXShop was extended with XeTeX support.

The Tiger version of TeXShop is capable of jumping from preview to code and vice versa without `pdfsync.sty`, using the PDF search technology built into Tiger.

Starting with version 2.18, TeXShop has included support for SyncTeX. This technology also allows jumping from preview to code and vice versa without including any special style file, but is much more reliable than PDF search, especially for documents that include mathematical formulae.

As always, we caution you that the hot software of today is the forgotten relic of tomorrow. You need to keep yourself aware of how the technology is developing.

## 6.4   Spell-Checkers, Grammar Checkers, and the Like

### 6.4.1   Spell-Checkers

The great thing about a document created on a computer is that the document is stored on your hard disk as a computer file. Thus your document has become a sequence of bytes. In most cases, your document in electronic form will consist primarily of `ASCII` code—`ASCII` (American Standard Code for Information Interchange) is the international code for describing the characters that appear on your computer keyboard and your computer screen. The computer can analyze these bytes in a number of useful ways. One of the most useful of these is the spell-checker.

At this writing, spell-checkers are highly sophisticated tools. A good spell-checker can zip through a 10,000-word document of ordinary text in a few minutes. It will flag a word that it does not recognize, suggest alternatives, and ask you what you want to do about it. It will catch many standard typographical errors, such as typing "naet" for "neat," or such as typing the word "the" at the end of line $n$ and also at the beginning of line $(n + 1)$. Of course it will also flag most proper names, archaic spellings, and many foreign words and mathematical terms. As you use your spell-checker, you can augment its vocabulary (which is performed semiautomatically, so requires little labor), hence your spell-checker becomes more and more accustomed to *your particular writing*. Given that a spell-checker requires very little effort to learn and use, and that it can only add to the precision of your document (it suggests changes, and makes only those that you approve), you would be foolish not to use a spell-checker. *However:* Never allow the spell-checker to lull you into a false sense of security. To wit, the ultimate responsibility for correct spelling lies with you (see below for more on the limitations of spell-checkers).

If you use a garden-variety spell-checker on a TEX document, then you will be most unhappy. The spell-checker will flag every TEX command (words beginning with \ ) and every math formula (set off by $ signs). You will find processing even a short TEX document to be an agony. Today the text editors that most TEX aficionados will use (such as `WinEdt`) have a built-in spell-checker that knows TEX.

Do not use a spell-checker foolishly. If you intend to write the word "unclear" and instead write "ucnlear" (a common transposition error), then the spell-checker will certainly tell you, and this is useful information. But if you intend "unclear" and instead write "nuclear," then the spell-checker will forge blithely ahead—because "nuclear" is a *word*, and a spell-checker will only flag non-words. If you mean to say that someone is "weird" and instead you say he/she is "wired," then your message may still trickle through; but your spell-checker will not help you to get it right. The lesson is clear (rather than unclear): if your document passes the spell-checker, then you know that certain rudimentary errors are not present; however, certain other, more sophisticated, errors could be present. Will you have to catch them yourself, with old-fashioned proofreading? Read on.

### 6.4.2 Misuse of Spell-Checkers

One of the big events in the world of finance in many years is the invention of the Black/Scholes option pricing scheme. This very sophisticated technology uses stochastic integrals—*very* advanced mathematics. It won the Nobel Prize for Myron Scholes (Fischer Black had died). Naturally Scholes's school, Stanford University, wanted to make a big deal out of its distinguished faculty member's encomium, so an article was written for the in-house newsletter. Unfortunately, some foolish editor ran a spell-checker on the article and ended up changing every occurrence of "Myron Scholes" to "moron schools." And this is the way that the article appeared. Woe is us.

We conclude with another anecdote, courtesy of G. B. Folland. One of Folland's publishers used a spell-checker that recognized the word "homomorphism" but not the word "homeomorphism." The result? The copy editor changed every instance of the latter to the former. The original manuscript contained several dozen of each. Now do you see how a spell-checker can get a person into trouble?

The bottom line is that you yourself are ultimately responsible for the accuracy of your manuscript. Use a spell-checker by all means, but then double check that spell-checker.

## 6.5   What Is TEX and Why Should You Use It?

### 6.5.1   The Revelation that Is TEX

TEX, created by Donald Knuth in the early 1980s, is an electronic typesetting system. Designed by a mathematician, specifically for the creation of mathematical documents, it also is a versatile tool in other typesetting tasks. In fact TEX is used in many law offices, and was once used to typeset *TV Guide*. It is particularly useful for typesetting foreign languages like Greek and Russian. The reference [Kn] tells something of the philosophy behind the creation of TEX.

What makes TEX such a powerful tool? First, TEX is almost infinitely portable. A TEX document created with an Apple computer in Hong Kong can be sent over email to a PC user in Sheboygan who in turn can send it on to the user of a Cray I in Bielefeld. During this process, there are never any problems with compiling, printing, or viewing.

The book [SaK] already contains this author's efforts at describing the inner workings of TEX and how to learn them. I shall not repeat that material here. Instead, I shall say just a few words about how TEX is used.

### 6.5.2   TEX is Not a Word Processor

TEX is *not* a word processor. Instead TEX is what is called a "markup language." "Markup language" means that, in your TEX document (created with a text editor—not a word processor), you enter commands that tell TEX what you wish to have appear on each page, and in what position. TEX allows you to position each character on the page to within $10^{-6}$-inch accuracy.

If you think about all the material that appears on a typeset page, then what is described in the last paragraph sounds arduous—like it is simply too much trouble. Fortunately, TEX performs most typesetting tasks and decisions automatically for you.

### 6.5.3   Typesetting Ordinary Prose

If you are typesetting ordinary prose, then you simply type the words on the screen, with spaces between consecutive words. With TEX, you can leave any amount of space between successive words in your source code; you can also put any number of words on each line of code. TEX will choose the correct spacing, and the correct number of words for each line, when it compiles the document. You indicate a new paragraph by leaving a blank line. There is almost nothing more to say about typesetting text: TEX spaces letters correctly, it chooses the correct amount of space to put between words, puts space between paragraphs, and so forth. It makes each line come out flush right, and ensures that each page contains the correct number of lines—not too many and not too few.

For mathematics, there are English-language-like commands that tell TEX just what you want. I will present just one example: The code would typeset as

```
Now it is time to do some mathematics---a task for
which, given that we have spent many years at the
university, we are eminently well prepared.  Our work
is inspired by the identity $x(1 + x) = x + x^2$.

Let us consider the equation
\[
\int_x^{x^2 - x} {{\alpha^3
   + 17{{\alpha} \over {\alpha-2}}} \over
   {{\alpha-5} \over {\alpha + 1}}} \, d\alpha
    = \operatorname{det} \,
   \left (
   \begin{array}{ccc}
   x^2 & 3x & x \\
  {{x^3 - 4} \over {x + 1}} & \sin x & \log x \\
  {{x}  \over {x+1}} & \operatorname{erf}\, x & \sqrt{x} \\
   \end{array}
   \right )
\]
which has been a matter of great interest in recent years.
```

Now it is time to do some mathematics—a task for which, given that we have spent many years at the university, we are eminently well prepared. Our work is inspired by the identity $x(1+x) = x+x^2$. Let us consider the equation

$$\int_x^{x^2-x} \frac{\alpha^3 + 17\frac{\alpha}{\alpha-2}}{\frac{\alpha-5}{\alpha+1}}\, d\alpha = \det \begin{pmatrix} x^2 & 3x & x \\ \frac{x^3-4}{x+1} & \sin x & \log x \\ \frac{x}{x+1} & \operatorname{erf} x & \sqrt{x} \end{pmatrix}$$

which has been a matter of great interest in recent years.

Even though you may not know TEX, you should have little difficulty seeing the correspondence between the code entered and the resulting output. (Note that the single dollar signs signify material to be typeset in "in-text" math mode; the double dollar signs tell TEX first to enter, and then to exit, "displayed" math mode.)

### 6.5.4   Compiling a TEX File

After you have created an ASCII file with your text editor, call it `myfile.tex`, then, depending on which TEX format you used, you compile it with a command like `tex myfile` (for plain TEX), `latex myfile` or `amstex myfile`. Or, if you're using LATEX, you might use a wrapper such as `texify` or `latexmk`, which can automatically process indexes and bibliographies and run LATEX multiple times to resolve cross-references. This creates the "device independent file," called `myfile.dvi`. The `dvi` file can be ported to a printer, to a screen, or translated to `Postscript` or `*.pdf`. Modern versions of TEX can optionally generate a PDF file directly.

As you can see from the preceding example, TEX does a magnificent job of typesetting mathematics. Usually no human intervention is required in order to obtain the quality and precision that you desire.

### 6.5.5   Do Not Use a Word Processor

I should stress that you should *not* use a word processor to create your TEX source file. A word processor has a very large number of hidden binary commands for formatting and visualizing. These hidden commands will confuse the TEX compiler, and give you output quite different from what you want. You will get a lot of confusing error messages. You should instead use a text editor or another tool for inputting computer code. See Section 6.3 on text editors to find out what a text editor is and how to get one.

### 6.5.6   TEX Preview

One interesting feature of TEX is that you cannot expect to see on the screen exactly what you will obtain in your printed output. For even a high-quality screen has resolution about 400 or so pixels per inch. Today, printers have a resolution of 2400 or more dots per inch. The `Preview` programs that come with TEX allow you to view your document to the extent of seeing where the various elements appear on the page—sufficient for doing elementary editing. But, to view the final output accurately, you must print a hard copy.

### 6.5.7 Graphics in TEX

It is fairly straightforward to import a graphic into a TEX document. An example of the relevant command is

```
\begin{figure}
\centering
\includegraphics[height=2.25in, width=2.75in]{m:/books/fig6-2.eps}
\caption{A ``pop'' or ``click.''}
\end{figure}
```

You can see that we use the `\begin{figure}`-`\end{figure}` command to create a "float." This means that the figure does not have a fixed position but is floated around for a best fit. The `\includegraphics` command specifies height and width for the figure, and also calls in the specific graphics file using a path statement. The `\centering` command does left-right centering of the figure. There is also a caption command.

Be sure to put this line in the preamble of your TEX document (the preamble is the material that appears before the `\begin{document}` command):

```
\usepackage{graphicx}
```

This will ensure that TEX can understand your graphics commands.

### 6.5.8 The Flexibility of TEX

TEX was originally designed with the notion of maximum power and flexibility in mind; Knuth planned that each discipline would develop its own style files to tailor TEX to its own uses. The variant LATEX, created by Leslie Lamport, endeavors to serve all end users. More specialized style files are available from the American Mathematical Society (to give just one example); these enable the AMS-LATEX user to typeset a paper in the style of any of the AMS primary journals.

There is a whole new world of document-preparation tools available today. As a semi-neanderthal, I would be more than sympathetic if you do not want to dive into all the graphic and typesetting and electronic features that I have described here. In fact these tools are best learned in gradual stages. The learning curve for TEX alone is rather steep, although the book [SaK] makes strides toward jump-starting the learning process. My recommendation is to begin by learning some form of TEX. LATEX is a particularly popular form of TEX, and one favored by publishers (because it is more structured and steers the author toward more standard formatting styles than does Plain TEX). The reference [SaK] creates an accessible bridge between Plain TEX (the most flexible TEX tool) and LATEX (the least flexible TEX tool). Most mathematics departments have the hardware, the software, and the expertise to make it easy for you to learn TEX. This software is one of today's standard mathematical tools. You are shooting yourself in the foot not to learn it.

### 6.5.9   Creating Graphics

For graphics, you may find that Adobe `Illustrator` or Corel `DRAW!` or (on a `UNIX` system) `Inkscape`[9] (which is just one of many `SVG` editors) is a useful utility. It may be noted that `Inkscape` replaces `xfig`. An even more recently developed, and really cutting edge, utility is `TikZ`.[10]

Any of these devices will output graphics in `*.pdf` format or `*.bmp` format or `*.eps` format or dozens of other popular graphics formats.

### 6.5.10   Front Ends for TEX

There are various front ends available to make TEX more user friendly. On a `Windows` machine, `Writer2LaTeX` is a popular choice. On a Mac machine, many people like TEXShop. For myself, LATEX is sufficiently friendly. I can work with it comfortably.

---

[9]`Inkscape` is a free and open-source vector graphics editor; it can be used to create or edit vector graphics such as illustrations, diagrams, line arts, charts, logos, and complex paintings. `Inkscape`'s primary vector graphics format is Scalable Vector Graphics (SVG); however, many other formats can be imported and exported.

   `Inkscape` can render primitive vector shapes (e.g., rectangles, ellipses, polygons, arcs, spirals, stars, and 3D boxes) and text. These objects may be filled with solid colors, patterns, radial, or linear color gradients and their borders may be stroked, both with adjustable transparency. Embedding and optional tracing of raster graphics is also supported, enabling the editor to create vector graphics from photos and other raster sources. Created shapes can be further manipulated with transformations, such as moving, rotating, scaling and skewing.

[10]`PGF/TikZ` is a tandem of languages for producing vector graphics from a geometric/algebraic description. `PGF` is a lower-level language, while `TikZ` is a set of higher-level macros that use `PGF`. The top-level `PGF` and `TikZ` commands are invoked as TEX macros, but in contrast with `PSTricks`, the `PGF/TikZ` graphics themselves are described in a language that resembles `MetaPost`. Till Tantau is the designer of these languages, and he is also the main developer of the only known interpreter for `PGF` and `TikZ`, which is written in TEX. `PGF` is an acronym for "Portable Graphics Format." `TikZ` was introduced in version 0.95 of `PGF`, and it is a recursive acronym for "TikZ ist kein Zeichenprogramm" (German for "TikZ is not a drawing program").

   The `PGF/TikZ` interpreter can be used from the popular LATEX and ConTEXt macro packages, and also directly from the original TEX. Since TEX itself is not concerned with graphics, the interpreter supports multiple TEX output backends: `dvips`, `dvipdfm/dvipdfmx/xdvipdfmx`, `TeX4ht`, and `pdftex`'s internal `pdf` output driver. Unlike `PStricks`, `PGF` can thus directly produce either `PostScript` or `pdf` output, but it cannot use some of the more advanced `PostScript` programming features that `PStricks` can use due to the "least common denominator" effect. `PGF/TikZ` comes with extensive documentation.

There are a number of implementations and distributions of TEX. The TEX User's Group homepage (`www.tug.org`) contains links to all of the major free distributions, such as TEX Live and MikTEX. Another possibility is the commercial PCTEX product. All are very good.

## 6.5.11 Different TEX Outputs

Knuth originally designed TEX to output a `*.dvi` file. Here "dvi" stands for "device independent." In the old days, one had a separate piece of software for converting the `*.dvi` file to a `*.hp` file that an HP printer could understand. In my original TEX installation (around 1988) it took 3.0 minutes per page to compile a TEX file and 3.5 minutes per page to convert the `*.dvi` file to an `*.hp` file. So, for instance, to produce a hard copy of a chapter of a book took a couple of hours! These days I can do the job in just a few seconds. Now I can compile a 500-page book in about 3 seconds, and convert the `*.dvi` file to a `*.pdf` file in another 3 seconds.

In general today things are more streamlined. Working in MikTEX, it is quite common for a user to compile the file `myfile.tex` with the command

```
pdflatex myfile
```

This directly produces a `*.pdf` file as output. Most printers these days can understand `*.pdf` files, and can print them easily. And you can view the `*.pdf` file with an Adobe reader, downloadable for free from the Internet. Products like `TeXShop` will produce a `*.pdf` file directly from one's LATEX file, without any intermediate steps.

If you are a dedicated `Mathematica` user, then you may know that it is possible to create a TEX document *inside* `Mathematica`. Indeed, `Mathematica` can emulate LATEX.

`LyX` is an open source document processor based on top of the LATEX typesetting system. Unlike most word processors, which follow the `WYSIWYG` paradigm, `LyX` has a `WYSIWYM` ("what you see is what you mean") approach, where what shows up on the screen is only an approximation of what will show up on the page.

Since `LyX` largely functions as a front end to the LATEX typesetting system, it has the power and flexibility of LATEX, and can handle documents including books, notes, theses, academic papers, letters, etc. Knowledge of the LATEX markup language is not necessary for basic usage, although a variety of specialized formatting is only possible by adding LATEX directives directly into the page.

`LyX` is popular among technical authors and scientists for its advanced mathematical modes, though it is increasingly used by non-mathematically-oriented scholars as well for its bibliographic database integration and ability to manage multiple files. `LyX` has also become popular among self-publishers.

As previously noted, Donald Knuth did not market TEX. In fact he *gave* the copyright to *The TEXbook* and the trademark on the "TEX" name to the

American Mathematical Society. The AMS conscripted Michael Spivak to create AMSTEX macros, which enhanced plain TEX's rather meager facilities for document structure and advanced mathematical markup with a more sophisticated set of standard macros, making it easier for authors to collaborate and for publishers to reuse author-prepared files. Once it became clear that Leslie Lamport's LATEX provided a superior framework for creating structured documents and incorporating extensions in the form of document classes and packages, the mathematical features of AMSTEX were rewritten and ported to LATEX as the popular `amsmath` package. This is now part of the core LATEX support for mathematics. Along with `amsfonts`, `amsrefs`, and document classes such as `amsart` and `amsmath` form the collection of extensions to LATEX known as AMS-LATEX. Today most mathematicians, and most publishers, use LATEX; it has become the *lingua franca* of the TEX world.

## 6.5.12   Spivak's Fonts

When Donald Knuth invented TEX, he also created the Computer Modern font (using his utility `MetaFont`). This is a simple and elegant font. Most mathematical TEX users are content to typeset their work either using Computer Modern or Times Roman.

Michael Spivak invested a great deal of time and money and effort in developing a new set of fonts (the MathTime Professional fonts) that are an alternative to the Computer Modern font of Knuth. Spivak's observations included that

- Some of the Greek letters in Computer Modern were hard to distinguish from others.

- The large parentheses in Computer Modern are not rounded as they should be.

- The root signs in Computer Modern are not designed properly.

There are a number of other technical ways in which the MathTime fonts are an improvement over the Computer Modern. Spivak's fonts may be purchased on the Internet. There is a trial version of the fonts available for free.

## 6.5.13   LATEX Macros for Tagging

LATEX has particularly useful macros for tagging formulas and theorems and for creating bibliographies. In fact LATEX lets you assign a *nickname* to each formula and each theorem. When you later refer to the formula or theorem, you refer to it by nickname. Then, when you compile the TEX file, everything is numbered correctly. Even if you have moved all your formulas and theorems around, LATEX is smart enough to get all the numbering right. Similarly for the bibliography: you give each reference a nickname, and when you refer to a reference you do so by nickname. You can choose from among many different formats for the Bibliography. When you compile, everything comes out just as it should.

BibTeX also offers you the option of creating a bibliographic database. When writing your paper, you call in references from this database (usually by using an assigned nickname). Over time, your database grows and may eventually have thousands of books and papers and other items. It is a great resource when you are creating a large bibliography.

The world of TeX has become a way of life for many people. The TeX Users Group (or TUG) is an organization dedicated to promoting and developing TeX. The TUG newsletter is a fascinating read for those interested in TeX.

### 6.5.14 Books About TeX

There are a number of good books that will acquaint you with the world of TeX. Among these are the TeXBook by Knuth [KnT], several books by Grätzer [Gra1], [Gra2], [Gra3], the *LaTeX Companion* [Goo], and the book of Sawyer/Krantz [SaK].

With TeX or LaTeX, you yourself do the typesetting of your work. So, when you send the work in to a publisher, the typesetting should be considered a done deal. Right? Wrong.

### 6.5.15 TeX and the Production Department

The trouble is that many publishers have their own production group that wants to make sure that all articles in a given issue of a journal, or all chapters in a given book, are typeset in the same way and in the same style. We want all the bibliographic references to be formatted in the same way, and all the equations to be displayed in the same way, and so forth. The problem with this is that the typesetters do not know any mathematics. Well meaning though they are, they can introduce spurious line breaks and other artifacts and, as a result, alter the meaning of an equation. Sometimes with disastrous effects.

Any good journal or book publisher will give you the opportunity to proof-read your article *after* the in-house editors have worked their magic. Do take that opportunity to doublecheck everything and to ensure that your work comes out the way that it should.

### 6.5.16 TeX Macros

A concluding thought about TeX is this. Once you really get into this new tool, you will learn how to create your own macros, and you may even use PS (Photoshop) or OTF (OpenType) fonts. Thus your TeX source code file will contain many "calls" to ancillary files. That may be fine for you, but publishers hate it. They prefer to receive a single TeX source code file from you that will compile all by itself. And certainly if you are going to be posting your work on a preprint server like `arXiv` (see Section 7.1), then you want your source code file to be self-contained. Give this matter some thought as your TeX skills develop.

We conclude this section with an amusing story about TeX. Mathematician Pete Casazza of the University of Missouri wrote a book several years ago.

Naturally he wrote it in TEX. At some point in the publication process, his publisher sent him a snail mail package containing edited page proofs and a disk with the TEX file on it. Now the University of Missouri is in Columbia, Missouri. But the U. S. Postal Service sent the package to the *country* of Colombia. Well, the authorities in Colombia opened the package and were most curious as to what was on the disk. They examined the TEX file and saw all the dollar signs (recall that dollar signs are used to format math formulas). They rapidly concluded that these were the ledgers for an illegal drug cartel. After a few months of hard study, they were unable to determine just what the file was telling them. So they enlisted the help of the FBI. The FBI was able to discover that this was in fact a TEX file containing nothing but mathematics. So, after a delay of a good many months, Pete Casazza finally got his package and was able to proceed with the publication of his book.

## 6.6   Graphics

### 6.6.1   Creating Graphics

As indicated elsewhere, a common (though somewhat antiquated) method for including graphics in a book is still to create them *separately*, each on its own page. The drawings could be created by hand, with pen and ink. Or they could be produced with Corel DRAW!®, or MacDraw®, or Adobe Illustrator®, or Inkscape, or TikZ, or any number of other packages. To repeat, each figure should be on a separate piece of high quality drawing paper (available from any store that carries art supplies) and drawn in dense black ink. Use a proper drawing pen—not a ball point, or a rolling writer, or a pencil. Best is to draw the figures (considerably) larger than they will actually appear in the book, in thick dark strokes. When they are photographically reduced to fit, then the pen strokes come out sharper, denser, and darker.

Each figure should be labeled clearly: a typical label might be

**Chapter 3     Section 2     Figure 5**

Correspondingly, somewhere in Section 2 of Chapter 3 there should be a space set aside for this figure, and it should be labeled "Figure 5." And be sure that the text contains a specific reference to each figure by name (label or caption); do not leave it to the reader to determine what figure goes with which set of ideas. (The same remark applies, of course, to tables.) It helps, though it is not mandatory, to give each figure a caption.

Drawing good illustrations for your work is an art. A good figure is not too busy, does not have extraneous information or extraneous penstrokes, and displays its message prominently and clearly. The books [Tuf1] and [Tuf2] by Edward Tufte will give you a number of useful pointers on how to develop powerful graphics for your work.

### 6.6.2 Ideas of Edward Tufte

Of course we all know that there are copious electronic tools for creating art-work in your manuscript. Just as an example, many versions of TeX have simple commands, such as \includegraphics, that allow you to import an encapsulated Postscript file into your document. In one common scenario, a \special command insets raw printer commands into the file that will communicate with your printer. The result is that your Postscript figure appears right on the printed page (*provided* that you have a Postscript printer or know how to use Ghostscript to make Postscript talk to a non-Postscript printer). Some versions of TeX—such as Personal TeX®—understand several other graphics languages as well. Many graphics programs give you a choice of several different graphics output languages; these could include ps, eps, pdf, bmp, or wmf graphics images. The documentation for your TeX software (for instance Personal TeX) will explain precisely which graphics languages it can handle and how it does so.

### 6.6.3 Symbol Manipulation Software

And now a caveat about Mathematica®, Maple®, and the like. These, too, are small miracles. If you need to draw a hyperboloid of one sheet, or the graph of $z = \log(|\sin(x^2 + y^3)|)$, then there is nothing to beat Mathematica. I recommend that you use it. Mathematica will output your figure in encapsulated Postscript, for storage on your hard disc, and in principle this file can be imported into your document.

A final note: ask Mathematica to graph a horrendously complicated function of two variables, and it will do so in an instant. Such tasks are what Mathematica is designed to perform. And it will provide the labels on the axes automatically. But endeavor to draw a rectangle or triangle, and to label the vertices in your own fashion, and it may take you an hour. Conversely, I can hand draw the rectangle or triangle and provide the labels in five minutes. But it could take me hours to graph the function. Instead you should draw the triangle or rectangle using Corel DRAW! or Adobe Illustrator. *Use the proper tools in the proper context.*

## 6.7 The Internet and Electronic Journals

### 6.7.1 What About Electronic Publishing?

Just a word about electronic publishing in general. The spirit of electronic publishing is to bypass the traditional hard copy of published materials, and instead make the materials available on the Internet. Readers would be identified and would pay either by buying a password or by paying the publisher to make materials available to a *particular* CPU with a particular identification number (the IP address—given by four octets of code).

## 6.7.2   Hypertext

A part of this new electronic publishing environment is `hypertext`. With `hypertext`, certain words or phrases in the electronic document appear in an accented form—often in a different color or underlined. If the reader "clicks" on the accented word, then he/she is "jumped" to a cognate item on a different web page. For instance, if you are reading a book on the function theory of several complex variables, you come across the word "pseudoconvex," and you cannot recall what it means, then—instead of madly flipping through the book trying to find the definition (this is the old way)—you click on the word and are jumped either to the passage that contains the definition, or perhaps to a lexicon, or perhaps to a menu that offers you several options. Alternatively, you could click on a reference to another book or paper and you would be jumped to the reference—to the *actual text of the reference*—no matter where in the world the source is.

## 6.7.3   Electronic Journals

There now exist many electronic journals. An electronic journal is one in which all transactions—submission, remanding to a referee, referee's report, editorial decision, and publication—are executed over the Internet. No hard copies of the journal are produced, nor archived.

Several of these new electronic journals are "startup" journals, run "for love" by an individual from his/her office computer. Others are institutionalized, but are still free. Still others are produced by commercial publishers and require a paid subscription.

## 6.7.4   Open Access

Begun in the 1990s, the notion of *Open Access* has begun to play a major role in journal publishing—not just in mathematics but across the sciences. There are now over 10,000 Open Access journals. Open access journals have been heavily promoted by NIH (the National Institutes of Health). An Open Access journal makes its contents available to the world at large for no charge.

Nobel Laureate Harold Varmus was the Director of the NIH from 1993 to 1999. He used that position in part to exert a notable influence on scientific publishing. Varmus is no fan of commercial research journals, so he helped to jump start the Open Access movement.

Near the end of his tenure as NIH director, Varmus became a champion of ways to more effectively use the Internet to enhance access to scientific papers. The first practical outcome was the establishment, with David Lipman of the National Center for Biotechnology Information at NIH, of *PubMed Central*, a public digital library of full-length scientific reports; in 2007, Congress directed NIH to ensure that all reports of work supported by the NIH appear in *PubMed Central* within a year after publication (see [Vas] and [Pub]). Varmus and two colleagues, Patrick Brown at Stanford and Michael Eisen at U.C. Berkeley, were co-founders and leaders of the board of directors of the Public Library of Science (PLoS), a not-for-profit publisher of a suite of open access journals in the biomedical sciences.

It is now the case that scientists who receive NIH funding are *required* to make their work available to the public after a certain amount of time. This can be done via an Open Access journal or, for instance, by posting the paper in *PubMed Central*. The National Science Foundation is developing a website for Open Access analogous to NIH's *PubMed Central*.

Just what is Open Access (OA)? The basic principle of an OA journal is that anyone (whether affiliated with a university or not) can read the articles in the journal without paying for a subscription or any fee at all. There are typically no subscriptions to an OA journal. This is fine for a journal run by a group of volunteers on a personal computer. But if a commercial publisher chooses to publish an OA journal, then someone has to pay the publisher's expenses, and the publisher needs to make a profit. So a fee (an *article processing charge*, or APC) is levied against the author, and the fee is generally nontrivial. Often the fee is in the thousands of dollars. The largest open-access publishers—BioMed Central and PLoS—charge \$1,350–\$2,250 to publish peer-reviewed articles in many of their journals, although their most selective offerings charge \$2,700–\$2,900. A paper that costs US\$5,000 for an author to publish in *Cell Reports*, for example, might cost just \$1,350 to publish in *PLoS ONE*, whereas *PeerJ* offers to publish an unlimited number of papers per author for a one-time fee of \$299.

Proponents of OA hope that the model for journals will shift: from the university paying many thousands of dollars per year in subscriptions to instead the university paying many thousands of dollars per year in APC charges. Some universities have in fact bought in to this new model. Most have not. In some cases the author will have a grant that will pay the fee. In most cases (especially in mathematics) this is not so. Some (see [Ewi]) have argued that OA is turning scholarly journals into vanity presses. Others have observed that dishonest publishers will accept a great many substandard papers just to generate a strong cash flow.

The main point here is that the Internet has opened up a world of new possibilities for scholarly publishing. OA is just one of these, and it is something that we all need to learn to live with. The article [Ewi] presents a cogent analysis of the OA movement.

There are different flavors of Open Access, and a good place to read about the details is [Sub]. OA delivered by journals is called *gold*. For gold OA, the author typically pays an APC and the article becomes freely available to the public immediately. OA delivered by repositories is called *green*. A green OA article becomes available to the public after one year. *Libre* OA involves removing permissions. The Open Access Newsletter, created by Peter Suber, will give you the chapter and verse on OA from the point of view of its partisans. Its URL is

```
http://legacy.earlham.edu/~peters/fos/newsletter/archive.htm
```

Some hard-copy journals are now simultaneously publishing an electronic version. One interesting innovation is that some traditional journals make any mathematical paper available electronically, for a modest charge, as soon as the paper has been accepted.

### 6.7.5 Electronic-Only Journals

There are several advantages of electronic-only journals: **(1)** the journals take no shelf space (a fact of immense importance to librarians), **(2)** the journals cannot be lost or stolen, **(3)** an unlimited number of readers can access any given article at the same time, **(4)** (in many cases) individuals can print out their own hard copies of any given article.

What are the disadvantages of electronic journals? The most significant of these is the archiving issue. With a hard-copy journal, a thousand hard copies go to a thousand libraries all over the world. The chances that all one thousand of these will be destroyed are slight. We can still read books and journals that were written 2000 years ago, and we do. It only takes a pair of eyes.

With an electronic journal, there are only electonic copies and electronic backups. Sun spots could wipe out all of these in a jiffy. There is also the question of having hardware available to read the electronic files. Today any file created on a `Windows` machine can be read by any other `Windows` machine. One hundred years from now, `Windows` will no longer exist. We will probably have completely different means of processing and accessing data. Optical disks will no longer exist, flash drives will no longer exist. Tape drives will probably no longer exist. Whatever means we use to store the electronic files today will be unheard of in 100 years. There are people—professionals—thinking about these questions. But most of us are not.

The world of electronic publishing is developing rapidly, and promises new frontiers of publishing activity and also of legal complications. As an example, the copyright law issues connected with electronic publishing are immense [Oke]. See also Section 4.7.

### 6.7.6 Free Books on the Internet

Some authors are making entire books, including textbooks, available at no charge on the Internet. A sample is the notable calculus book by Neal Koblitz that is available at

```
http://www.freetechbooks.com/california-free-digital-
calculus-textbook-t781.html
```

The website

```
http://people.math.gatech.edu/~cain/textbooks/onlinebooks.html
```

lists 77 math texts, some by rather distinguished authors, that are free online.

Commercial publishers are also rapidly developing the publication of electronic forms of books. Some publishers will offer a few sample chapters for free on the web; but if you want the whole book, then you have to pay.

In fact some publishers will propose to an author that a home page be set up for his/her book, and that not simply the book but also a variety of ancillaries appear on the website. These ancillaries could include relevant papers, a bibliographic database, exercise books, lecture manuals—you name it. In some scenarios, a publisher may develop a version of a book to which readers may contribute interactively. Some books appear these days with no Bibliography—instead the reader is referred to a website for the references. Not everyone thinks that this is good scholarship, but this is the world that we live in.

## 6.8 Collaboration by Email, Uploading and Downloading

### 6.8.1 Email Collaboration

Writing a collaborative mathematical work is a source of great pleasure. It is especially fun when you use email as a tool. Entire chapters can be zapped around the world in an instant. You get immediate feedback on your ideas. In many (but not all) ways, collaborating by email is like having your partner in the office next door.

### 6.8.2 Doing Your Work in TEX

Most of us find it convenient to work on a local computer—either a PC or a Mac. You of course store all your files on that local computer—on a hard drive or a flash drive or some other mass storage device. Then you typically upload the files to the school's computer system—using `ftp` or some similar protocol—check your local computer gurus for the best way to do this. Once you have your files on the school's `UNIX` system, then it is a simple matter to attach them to an email and send them to your collaborator. Usually it is best to send both a `*.tex` and a `*.pdf` file. If there are graphics files, then you may want to send those as well. I generally find it convenient to create a `*.zip` file that contains all the files that I want to send. That way, my files are bullet-proofed, and I only have to send one attachment. `Gmail`, the email system provided by `Google`, makes it particularly easy to send email attachments.

If you wish, you can type comments at the beginning of, or in the middle of, the TEX document. If you precede each line of the comment material with a % symbol, then TEX will ignore those lines. I usually include general comments, set off by % marks, at the top of the TEX file. And, in the body of the text, I include specific comments about particular lines in the paper or book. I usually begin a comment with an identifying squib, something like

```
%%%!!%%% Dear John:
```

`%%%!!%%%` This entire paragraph is in error.

That way my collaborator will have no trouble finding and reading my remarks. You may even find it convenient to attach a date to each comment.

## 6.9   Mathematical Collaboration in Today's World

### 6.9.1   Mathematical Collaboration

In the old days—say 100 years ago—mathematical collaboration was relatively rare. There were only several hundred research mathematicians in the world, and each of these sat alone in his or her office and applied himself/herself to the proving of theorems. Hardy and Littlewood, who together wrote more than 100 papers, were certainly the exception. In those days, the only two modes of discourse were snail mail or meeting face-to-face. There was the telephone, but long-distance calls were considered to be prohibitively expensive.

In today's world more papers are written collaboratively than not. And there are so many devices to enable this collaboration. Certainly collaboration by email is quick and convenient. Drafts of papers, written in TeX or LaTeX, can easily be sent as email attachments. One can use `Skype` to have multi-hour rap sessions with a collaborator on the other side of the world—with no cost to anyone.

`FaceTime` is an Apple product that works on `iPhones`, tablets, and Apple computers. It allows you to speak to a friend or collaborator and see him/her at the same time. So, in principle, both of you could be writing on a blackboard or white board and each could see what the other is doing. Collaborating with `FaceTime` is almost the same as being in the same room together.

Of course attending conferences, workshops, and research institutes is a terrific way to hook up with people who have interests similar to your own. After you have established a working relationship with some of these people, then you can go back to your home institutions and communicate by one of the methods described above.

### 6.9.2   `DropBox`

A convenient tool in collaboration is `DropBox`. Available for free from the Internet, `DropBox` allows you easily to upload and download very large files, organize them in useful ways, and make them accessible to certain designated individuals. Put in other words, it is a fact that many email systems cannot handle very large attachments. `DropBox` is a file hosting service operated by the American company Dropbox, Inc., headquartered in San Francisco, California, that offers Cloud storage, file synchronization, personal Cloud, and client software. On the other hand, `gmail` has a utility called `Drive` that *does* allow you to have email attachments of any size. It does so by exploiting the Cloud.

I have had 66 mathematical co-authors and I can tell you rather definitely that nothing beats being in the same room with your collaborator and kicking ideas around at a blackboard or on a tablet. It is a matter of synergy and exchange of energy. It is what works.

Even so, I have many co-authors all over the world, several of whom I have never met, with whom I collaborate by email. It is a system that works well, and is a pleasure to be part of.

# Chapter 7

# The World of High-Tech Publishing

*Never spend more than a year on anything.*

Jeff Ullman

*The commonest thing is delightful if only one hides it.*

Oscar Wilde

*Not of the letter, but of the spirit: for the letter killeth, but the spirit giveth life.*

The Holy Bible, the New Testament
The Second Epistle of Paul the Apostle
to the Corinthians. Chapter 3, Verse 6.

*The road to hell is paved with works-in-progress.*

Philip Roth

*The road to hell is paved with adverbs.*

Stephen King

*Who wants to become a writer? And why? Because it's the answer to everything. It's the streaming reason for living. To note, to pin down, to build up, to create, to be astonished at nothing, to cherish the oddities, to let nothing go down the drain, to make something, to make a great flower out of life, even if it's a cactus.*

Enid Bagnold

*To gain your own voice, you have to forget about having it heard.*

Allen Ginsberg, WD

*Cheat your landlord if you can and must, but do not try to shortchange the Muse. It cannot be done. You can't fake quality any more than you can fake a good meal.*

William S. Burroughs

*All readers come to fiction as willing accomplices to your lies. Such is the basic goodwill contract made the moment we pick up a work of fiction.*

Steve Almond, WD

### 7.0.1   The Impact of the Internet on Publishing

Before the invention of movable type by Gutenberg in 1492, books were expensive (generally they were hand-transcribed as a codex) and rare and really only available to the priveleged few. After Gutenberg, the situation improved a bit, but books were still a luxury. Today many books are available for free on the web. Many encyclopedias (especially, but not exclusively, `Wikipedia`) are also freely available. Many `Google` tools are free. This truly is the information age.

The advent of the computer, and particularly of the Internet, has completely changed the face of modern mathematical publishing and mathematical writing. There are many new artifacts and features of this world. And many new forces at play. In this chapter we attempt to describe the key new components of our publishing life.

## 7.1   Mathematics on the Web

### 7.1.1   Posting Your Work on the Web

In the old days, when you wrote a math paper, you painstakingly wrote it by hand on $8.5'' \times 11''$ paper. Often you had to mark all your math symbols with one color of ink, and your Greek letters with another color of ink, and your fraktur letters with another color of ink. And hope and pray that the person performing the next step of the process would understand what you wanted.

Then it was typed up by a manuscript typist using an IBM Selectric typewriter. This typewriter was special because it had "element balls" with special characters such as math symbols, letters from the Greek alphabet, and special braces and brackets. It was still necessary to render some symbols by hand with an inkpen, but the Selectric did most of the work.

You would have many copies of this work reproduced on the photocopy machine, and you would mail these (with snail mail) to your colleagues all over the world. This is how a mathematician would establish his/her reputation and make his/her mark on the profession. You could not afford to wait for the paper to be published; this could cause a delay of a few years, and your likelihood of getting scooped was nontrivial. You had to get the word out right away. This is how it was done.[1]

The trouble with the system just described is that it meant that well-established people at the top universities heard all the new developments first. More obscure mathematicians who were not well connected were generally left out of the loop. They could go to conferences and get some hints about new developments. But they did not get their information in a timely fashion.

---

[1] Of course you could also give seminars and speak at conferences. This was an important part of the profession. It was also quite common to send out postcards announcing results. But electronic media were not at all available fifty years ago. We had only the postal system that was instituted in the United States by Benjamin Franklin more than 235 years ago. It should be noted that he was anticipated by a British system that goes back at least to Henry VIII and before that to the Persians.

## 7.1.2   Preprint Servers

Now things have changed. There are a good many preprint servers that serve as repositories for new mathematics. What is a preprint server? It is a website where you can post—as a `*.pdf` file or in some other electronic form—your new paper. Now we must understand clearly that a paper posted on a preprint server is not refereed or vetted in any way. It is just posted for all the world to see. And, indeed, absolutely anyone can view or download or print the papers posted on a preprint server. And virtually anyone can post on a preprint server (although some, like `arXiv`, have an entry level for submission). It is an observed fact that `arXiv` is the most popular and prevalent preprint server for mathematics. More will be said about this tool in what follows.

There are a number of specialized preprint servers for particular research areas of mathematics—for instance in $K$-theory and linear algebraic groups and mathematical physics and nonlinear science. But the most prominent and widely used mathematical preprint server is `arXiv`. Developed by Paul Ginsparg in 1991, `arXiv` started as a physics preprint server. But now it handles mathematics, computer science, statistics, quantitative finance, and quantitative biology as well. As many as 10,000 papers, in the six indicated fields, per month are posted on `arXiv`. It would be foolish to assert, as many people do, that "all math papers are now posted on `arXiv`." What is more accurate is to say that the number of papers posted on `arXiv` is 30% and growing. But there are plenty of older mathematicians who do not give a hoot about `arXiv`. And there are a number of other mathematicians who prefer to post their work on specialized preprint servers that are dedicated to particular areas of mathematics. And still others who just cannot be bothered.

## 7.1.3   `arXiv`

The website `http://www.arxiv.org` offers statistics on the use of `arXiv`. According to the latest data (as of 2017), there are 1,213,827 articles now posted on `arXiv`. If we estimate that 400,000 of these are in mathematics (a generous estimate), and we note that about 100,000 math articles are produced per year, and finally we note that `arXiv` is 25 years old, then it is easy to see that `arXiv` has not yet taken over. But it could.

Plenty of mathematicians are tired of dealing with obstinate referees and arrogant editors. They feel that, having posted their work on `arXiv`, they have published it and this is all that they owe to the mathematics profession.[2] One could argue the point. One could claim that the traditional refereeing and publishing process guarantees the robustness and longevity of our work. That traditional journals archive our work properly and reliably. That displaying mathematics as an undifferentiated melange of non-reviewed work is neither

---

[2]As an extreme instance, Grigori Perelman posted his Fields-Medal-winning work on the Poincaré conjecture and the geometrization program on `arXiv` and nowhere else. None of it has ever been published, even though top-notch journals contacted Perelman and asked him to publish his work with them.

productive nor useful. But these ideas are still very much in the air and still very much being debated.

The great thing about `arXiv` is that it is very easy for an end user to type in an author's name and get a listing of all his/her most recent papers. And it is equally easy to download any of them. You can also tell `arXiv` which areas of mathematics you are interested in, and it will send you an email notification each day of what new papers have been posted.

The server `arXiv` has become so well established that it is now possible, with many journals, to submit a paper by just providing a pointer to your `arXiv` posting. Most professional journals are fairly free and easy about `arXiv`. They will *not* insist that you take down your `arXiv` posting as soon as your paper is accepted by the journal. Book publishers are different, and they often *will* ask you to take your book down from `arXiv` once it is officially published. But of course book publishers are money-making organizations.

There are some mathematics journals today that work as follows. When you submit your paper, you simply tell the Editor the `arXiv` address of the work. The journal has the paper refereed. If the paper is accepted, then it appears in the next issue. And the next issue consists simply of a list of the titles and authors of accepted papers, together with a brief editorial introduction about each paper and a pointer to the `arXiv` address of each paper. These are called "`arXiv` overlay" journals, and they exist both in mathematics and in physics. An example of such a journal is Fields Medalist Tim Gowers's *Discrete Analysis* with web page

`http://discreteanalysisjournal.com`

This is what *Discrete Analysis* has to say about itself:

> Our articles live on the `arXiv`. This has a major advantage over a conventional journal—even if it is an electronic journal —which is that authors can post updates to their articles if they find ways of improving them. The link from the journal will always be to the accepted version, which will remain the version of record, but the associated `arXiv` page will notify readers if that version has been further updated. Thus, we have the best of both worlds: a permanent version of record, and also the possibility for authors to make subsequent improvements that readers will easily notice.
>
> Our website is properly designed for the internet age. It looks good, and one can extract useful information from it without an annoying amount of clicking and loading of new pages. That should not be a distinctive feature of *Discrete Analysis*, but if you go and look at the websites of other mathematics journals, you will quickly see the difference. As a bonus, the *Discrete Analysis* website displays well on your mobile phone.
>
> Articles published in *Discrete Analysis* are presented with "editorial introductions," which are written by the members of the editorial board and attempt to provide information (such as relevant

definitions, background context, and an idea of what is in the paper) that will help a reader browse the content of the journal in a pleasurable way. Sometimes we will find out this contextual information from referees' reports. A common complaint about the current peer-review system is that valuable information of this kind is seen only by the editors. With *Discrete Analysis*, readers get a chance to see some of it. (However, referees remain anonymous and the evaluative parts of referees' reports remain confidential.)

The esteemed magazine *Nature* published an article [Bal] about Gowers's efforts, and the website for that piece is

```
http://www.nature.com/news/leading-mathematician-launches
         -arxiv-overlay-journal-1.18351
```

An interesting feature of `arXiv` is that it only accepts TeX submissions. When you upload your paper, it must be in raw TeX or LaTeX form. Not `*.pdf`, not `*.docx`, not raw text. Only TeX. And `arXiv` *compiles* your paper right on the spot. If it succeeds, then you can proceed with the submission process. If it fails, then you are dead in the water. This is another motivation for you to learn to make your TeX files self-contained (see Section 6.5). In fact the `arXiv` website is quite explicit in stating that it prefers LaTeX2e. And, since this is the driving form of TeX behind MikTeX, and since MikTeX is currently the most popular and widely used form of TeX, that preference makes some sense.

If your paper has separate graphics files, then you may find it tricky to get `arXiv` to compile and accept your paper. But it can be done. I have done it.

As noted elsewhere in this book, once you write something, then it is immediately copyrighted to you. This is still true when you put a paper on `arXiv`. When you next submit the paper to a journal, it is quite standard for the journal to ask you to sign a copyright transfer agreement. When you sign it, then the copyright moves to the publisher.[3]

## 7.1.4 Posting Papers on `arXiv`

Just for the record, here is a fairly friendly journal publisher's policy toward posting papers on the `arXiv`:

> The ASL hereby grants to the Author the non-exclusive right to reproduce the Article, to create derivative works based upon the Article, and to distribute and display the Article and any such derivative work by any means and in any media, provided the provisions of

---

[3]Some mathematicians prefer to retain the copyright to themselves. This is because, for instance, it may happen years later that someone wants to put together a volume of historically influential papers in a certain subject area. If your paper is chosen for this volume, and if it is copyrighted to some other publisher, then nasty negotiations may ensue. And nasty fees. If you feel strongly about retaining the copyright to your work, you may have to negotiate with your journal publisher.

clause (3) below are met. The Author may sub-license any publisher or other third party to exercise those rights.

And here is a slightly less friendly policy, which still allows the author to post on `arXiv`:

> I understand that I retain or am hereby granted (without the need to obtain further permission) rights to use certain versions of the Article for certain scholarly purposes, as described and defined below (Retained Rights), and that no rights in patents, trademarks or other intellectual property rights are transferred to the journal.
>
> The Retained Rights include the right to use the Pre-print or Accepted Authors Manuscript for Personal Use, Internal Institutional Use and for Scholarly Posting; and the Published Journal Article for Personal Use and Internal Institutional Use.

### 7.1.5   Front **for the** `arXiv`

Now the truth is that `arXiv`, in its raw form, is rather stodgy and difficult to use. Fortunately for us, Greg Kuperberg has created a front end for `arXiv` called `Front`. The URL for `Front` is

`http://front.math.ucdavis.edu/`

`Front` is very user-friendly and easy to implement. I recommend it.

Of course the world, and especially the high-tech publishing world, is ever-changing. In twenty years, `arXiv` may no longer exist. But preprint servers have become a part of life. Some form of preprint server probably will still exist. And many of the points noted here about `arXiv` will apply in that new context. Twenty years ago the American Mathematical Society devoted a great deal of time and effort to developing a global preprint server for the mathematics community. But it got scooped by `arXiv`, and is no longer in use. This is how the world works. If you want your work to have some permanence, your best bet is probably to publish it in a journal of long standing and good repute. You should think of `arXiv`, or any other preprint server, as simply a device for getting the word out. It is *not* necessarily a device for archiving.

### 7.1.6   Many Versions of Your Paper

A nasty problem that we all have to deal with in the modern world is this. Once I write a paper, there are soon many versions of it floating around. There could be a dozen versions on my school computer, another dozen versions on my home computer, a version on `arXiv`, a version on my Home Page, and so forth. Which is the definitive version of the paper? There is no clear and easy answer to this question. You may want to give the matter some thought and establish a definitive policy for yourself.

## 7.2  MathSciNet

### 7.2.1  *Mathematical Reviews* online

We have mentioned `MathSciNet` at several earlier junctions in the book. Here we treat the topic with more care.

In 1869 Felix Müller and Carl Ohrtmann created the periodical *Jahrbuch über die Fortschritte der Mathematik*. Its purpose was to index the mathematical literature. The *Jahrbuch* was published by Walter de Gruyter in one volume per year until 1943. A total of 68 volumes, containing records of 200,000 publications, appeared in the *Jahrbuch*. One of the wonderful things about the mathematical literature is that it never goes out of date. Therefore, Bernd Wegner, Keith Dennis, and Elmar Mitter have created an online version of the *Jahrbuch* called `ERAM` (Electronic Research Archive for Mathematics) though most people refer to it as "The Jahrbuch Project" or the "JFM Project."

In 1939 Otto Neugebauer, who had in 1931 created *Zentralblatt für Mathematik* in Germany, fled from the Nazis and moved to Brown University in Providence, Rhode Island. There he created *Mathematical Reviews*. The purpose of both *Zentralblatt* and *Math Reviews* was to catalog and review the mathematical literature. Each of these journals publishes a brief review (a few paragraphs—rarely more than a page) about most of the papers published in most of the math journals around the world. Put in other words, each publication contains the bibliographical data of all recently published mathematical articles and books, together with reviews written by a community of experts from all over the world. In actuality, journals are classified by type: MR has a class of journals that they index cover-to-cover. For the others, each issue presents editorial decisions on the articles. Also, it is possible to include an item (article) in `MathSciNet` but not get a review for it.

When both *Math Reviews* and *Zentralblatt* were available in hard copy only, the two periodicals occupied equally hallowed positions in the firmament. Many math departments subscribed to both, and each had its partisans. But `MathSciNet` (see below) has now put the AMS's product in the forefront.

In 1980, *Math Reviews* was converted to an online database, and this eventually evolved into `MathSciNet` in the 1990s. It is safe to say that `MathSciNet` has revolutionized the mathematics profession. Now virtually any mathematician can, from virtually any location, look up papers and books in the mathematical sciences, assemble bibliographies and reading lists, and become acquainted with the literature. It is now relatively straightforward to assemble bibliographies and reference lists.

For the American Mathematical Society, *Math Reviews*/`MathSciNet` is a big enterprise. Situated in an old brewery in Ann Arbor, Michigan, at least 75 people are employed in the production of *Math Reviews*. A great deal of care is put into sorting out names (so that all the different John Smiths are distinguished), sorting out articles with similar titles, and getting all the bibliographic information correct. And a huge amount of effort is devoted to requisitioning and classifying and typesetting and organizing the reviews of the individual papers and books.

## 7.2.2   `MathSciNet` **and** `zbMATH`

The trouble with `MathSciNet` and `zbMATH` (see below) is that they are both
subscription based and quite expensive.[4] If you are situated at a university,
then you probably have access to one or both. If you are just working out of
your home, then you may not. But there are alternatives that are free. One
of these is `Google Scholar` with URL `https://scholar.google.com`. `Google
Scholar` is a freely accessible web search engine that indexes the full text or
metadata of scholarly literature across an array of publishing formats and disciplines. Released in beta form in November 2004, the `Google Scholar` index
includes most peer-reviewed online academic journals and books, conference papers, theses and dissertations, preprints, abstracts, technical reports, and other
scholarly literature, including court opinions and patents. It is not uncommon
these days for the Dean's tenure committee to want to know how many citations
a candidate has on `Google Scholar`. Another popular yardstick with deans is
the so-called *h-index* (named after Jorge E. Hirsch of U. C. San Diego, the inventor of the index). A mathematician's h-index is the greatest positive integer
$n$ so that this mathematician has published $n$ papers each of which has had $n$
citations. There are a number of indices of this nature; the i10-index is another.

## 7.2.3   Google Scholar

While `Google` does not publish the size of `Google Scholar`'s database (the
Google Scholar database encompasses all subjects, not just math and not just
science and technology.), third-party researchers have estimated it to contain
roughly 160 million documents as of May 2014 and an earlier statistical estimate
published in `PLOS ONE` using a mark and recapture method estimated approx-
imately 80-90% coverage of all articles published in English. `Google Scholar`
is freely available. It also, in some ways, resembles the subscription-based tools
known as Elsevier's `Scopus` and Thomson Reuters's `Web of Science`.

## 7.2.4   **Mathematics Indexing Services**

Unlike most other abstracting databases, `MathSciNet` takes care to identify authors properly. Its author search allows the user to find publications associated
with a given author record, even if multiple authors have exactly the same name.
*Mathematical Reviews* personnel will sometimes even contact authors directly to
ensure that the database has correctly attributed their papers. Author Search is
one of the three distinct ways a user can query `MathSciNet`. The other two are
Publication Search and Journal Search. For Publication Search, string match-
ing is used in all fields. This functioning is needed to allow for the database

---

[4]However, it must be noted that the American Mathematical Society is rather generous at
granting consortium pricing and special pricing on `MathSciNet` for groups of smaller colleges,
and also for institutions in third-world countries.

to access older reviews (pre 1985). However, users should be aware that many of the items from 1985 back through 1940 (MR's beginning) are not yet fully integrated. One of MR's main projects is to have all items from 1940 to the present fully loaded by the end of 2018.

We have noted that *Zentralblatt* also has an online product called `zbMATH`. `zbMATH` has an advantage over `MathSciNet` in that it includes all the electronic `ERAM` entries for the 200,000 articles that appeared in *Jahrbuch*. But today `zbMATH` is less prominent than `MathSciNet`. Neither *Math Reviews* nor *Zentralblatt* exists in hard copy anymore.

Of course `MathSciNet` and `zbMATH` rely on volunteers[5] to write the reviews of the mathematical articles that appear in their pages. It is a worthwhile activity to volunteer for such work. It does not take much time, and it makes you aware of papers that you otherwise might miss. I have gotten a few good ideas doing my `MathSciNet` work. In writing such a review, you often have to supply the Math Subject Classification Numbers for the article under review (if the author has not already done so). And you must generate some text about the paper. You need not wax loquacious here. At a minimum you should state, or at least paraphrase, the main results. You can, if you wish, comment on their significance. You also can, if you wish, remark on any shortcomings of the theorem, the proof, or the references. Some reviewers provide references to other relevant work (especially their own, although this is not encouraged). Some reviewers are quite critical, although I do not encourage that frame of mind—it does not add anything to the value of `MathSciNet` or `zbMATH` to have captious reviews.

All sorts of interesting searches can be done in `MathSciNet`. You can search for author(s), title, journal, MR number (which is a unique identifier for each item in the database), and many other choices for a paper or book that interests you. `MathSciNet` provides the user with the bibliographic information about an item in BibTEX format. `MathSciNet` uses `MathJax` to render mathematics in your browser so that you don't need to view the PDF version of a review in order for the mathematics to display properly.

You can also, on `MathSciNet`, calculate your collaboration distance to another mathematician. So, for instance, if you wrote a paper with Riemann who wrote a paper with Gauss, then your "Gauss number" is 2, and `MathSciNet` can calculate that for you.

If you look up `Steven Krantz` on `MathSciNet`, then it will tell you **(i)** how many papers and books Krantz has written, **(ii)** how many citations there have been of his work, and **(iii)** who his principal collaborators are. There is a wealth of information on `MathSciNet`, and this tool is certainly worth mastering.

---

[5]For each `MathSciNet` review, volunteer reviewers receive 12 AMS Points, which convert to ten dollars of credit that can be spent toward the purchase of AMS books.

# 7.3   Personal Web Pages

## 7.3.1   Your Personal Web Page

These days most every professional, most every politician, most every artist, and most every retail business has a web page. It is not difficult to set up a web page, and there is plenty of software available to help you do it.

My view is that your web page is a professional document and you should treat it as such. Some people have a button on their web page for personal data such as **(i)** spousal information, **(ii)** information about children, **(iii)** information about hobbies, **(iv)** a gallery of photographs, and so forth. I think it is better to have a professional web page for your professional activities and a completely separate web page—with a completely separate URL (Uniform Resource Locator, or web address)—for your personal activities. No sense to confuse the two.

You should take some care to lay out your web page so that it is easy for anyone to find the information they seek. This information could include an email address, mailing address, telephone number, FAX number, affiliated institution, your Curriculum Vitae, and so forth. Many people include information about courses taught, committees served, conferences organized, and other professional activities. That is fine.

The information should be laid out in tableau style (not essay style, or paragraph style). Careful use of color can help to highlight certain information. It is considered helpful and attractive to include a photo of yourself.

## 7.3.2   Using Word to Create a Web Page

One of the easiest ways to create a web page is to write it out in Microsoft Word. Then you can save what you have written as an HTML document. You will have to consult your local computer gurus to find out how to post this *.html document to actually create the web page, but this is just a technical detail.

## 7.3.3   Hyperlinks

Of course your web page can have hyperlinks to other web pages or documents. These other documents can include **(a)** your CV, **(b)** recent papers that you have written, **(c)** opinion pieces that you have written, **(d)** drafts of books that you have written, **(e)** a link to the AMS (American Mathematical Society) home page, **(f)** a link to the MAA (Mathematical Association of America) home page, **(g)** a link to the home page of your home institution, and so forth. Hyperlinks are part of what makes web pages useful and fun.

Many people have a link on their web page which, if a user clicks on it, he/she can immediately send an email to the person whose web page it is. Other people have a button that tells how many times the web page has been accessed.

Your web page is your face to the world, and you want it to reflect positively on you and what you do. Give it some serious thought as you create it. My web page has links to the web pages for the classes that I am teaching. So all my students see my web page. I want to give them all a good impression of myself.

## 7.4 Mathematical Blogs and Related Ideas

### 7.4.1 Blogs

A blog (a truncation of the word `weblog`) is a discussion or informational website published on the Internet consisting of discrete, often informal diary-style text entries (these are usually called "posts"). Posts are typically displayed in reverse chronological order, so that the most recent post appears first, at the top of the web page. Until 2009, blogs were usually the work of a single individual, occasionally of a small group, and often covered a single subject or topic. Since 2010, "multi-author blogs" have developed, with posts written by large numbers of authors and sometimes professionally edited. The rise of `Twitter` and other "microblogging" systems helps integrate multiple-author blogs and single-author blogs into the news media. The word "blog" can also be used as a verb, meaning to maintain or add content to a blog. The emergence and growth of blogs in the late 1990s coincided with the advent of web publishing tools that facilitated the posting of content by non-technical users who did not have much experience with `HTML` or computer programming. Previously, a knowledge of such technologies as `HTML` and File Transfer Protocol (`ftp`) had been required to publish content on the web, and as such, early web users tended to be hackers and computer enthusiasts. Since 2010, the majority of blogs are interactive web 2.0 web sites, allowing visitors to leave online comments and even message each other via GUI (Graphic User Interface) widgets on the blogs, and it is this interactivity that distinguishes them from other static websites. In that sense, blogging can be seen as a form of social networking service (although, in the next section, we are careful to distinguish blogs from social networking sites). Indeed, bloggers do not only produce content to post on their blogs, but also build social relations with their readers and other bloggers.

### 7.4.2 Blogs, Chat Rooms, and `Wikis`

By definition, a *blog* is a discussion or informational website published on the web and consisting of discrete, often informal, text entries (called *posts*). A *chat room*, by contrast, can have many contributors. A `Wiki` allows most anyone to *edit* the material being posted.

In mathematics, blogs have become very popular. See [Bae] by John Baez for an enthusiastic discussion of math blogs. John Baez's math blog can be accessed at `johncarlosbaez.wordpress.com`.

One particularly popular and successful mathematical blog is that created by Fields Medalist Terry Tao. It is full of information, not just about Tao but also about mathematics in general. Lots of pointers to interesting sites and interesting problems. One could benefit considerably by spending time with this blog. The URL is

```
https://terrytao.wordpress.com
```

A math blog can be about a particular mathematical topic, or even a single mathematical research problem. Participating in a math blog is very much like participating in a coffeeroom discussion, with many participants from all over the world. There is many a story of important problems being solved by the participants in a math blog or chat room.

### 7.4.3   Hundreds of Collaborators

Of course many new complexities can arise from a very large number of people working on a problem together. If 200 people contribute to the solution of a problem, then who writes it up? Whose name goes on the paper? Who decides where to submit it? If you submit to an Open Access journal and there is an (often nontrivial) author fee, then who pays it?

### 7.4.4   MathOverFlow

Some of the most famous and popular math blogs, chat rooms, and `Wikis` were created by Fields Medalists—notably Timothy Gowers and Terence Tao. Gowers has also created `polymath` and `MathOverFlow`. These are websites specifically designed to bring together large groups of people to work on specific mathematical research problems. `MathOverFlow` has had a number of notable successes.

It is particularly easy to participate in `MathOverFlow`. Go to `mathoverflow.net` and you will be immediately introduced to current problems under discussion. Further down the page you are asked to contribute your own comments. And so now you are a member of the gang!

### 7.4.5   polymath

It is equally easy to become involved in `polymath`. Just go to `https://polymathprojects.org/` and you are off and running. Also check out `http://michaelnielsen.org` for insights into `polymath`.

### 7.4.6   More on Blogs

One advantage of a blog is that it does not have to meet the usual scholarly publication standards, and does not have to fit into the purview, of any of the standard scholarly journals. It does not have to undergo any refereeing or vetting. You do not have to deal with tiresome referees or officious editors. You can create a blog about any topic that you think to be of interest, or that you think others will find to be of interest. This could include

(a) a discussion of an interesting paper that you read recently,

**(b)** how to come up with examples on your own,

**(c)** an attack on a specific research problem,

**(d)** partial results on a particular research problem,

**(e)** how to study for an exam,

**(f)** how to prepare for a job interview,

**(g)** your experience as a Project `NeXT` fellow,

**(h)** how to teach a certain unusual topic,

**(i)** how to handle tricky situations with students.

There are a number of websites that will help you to create a blog of your own. Among these are `siteblog`, `SiteBuilder`, `website`, `HostGater`, and `iPage`. These will help you to implement the sort of interactivity that makes a blog effective.

### 7.4.7 Chat Rooms

The term *chat room*, or chatroom, is primarily used to describe any form of synchronous conferencing, occasionally even asynchronous conferencing. The term can thus mean any technology ranging from real-time online chat and online interaction with strangers (e.g., online forums) to fully immersive graphical social environments. The primary use of a chat room is to share information via text with a group of other users. Generally speaking, the ability to converse with multiple people in the same conversation differentiates chat rooms from instant messaging programs, which are more typically designed for one-to-one communication. The users in a particular chat room are generally connected via a shared interest or other similar connection, and chat rooms exist to cater to a wide range of subjects. New technology has enabled the use of file sharing and webcam to be included in some programs. This would be considered a chat room.

### 7.4.8 The `Wiki` page

Another type of web page becoming very popular is the `wiki` page. A `wiki` is a website that provides collaborative modification of its content and structure directly from the web browser. In a typical `wiki`, text is written using a simplified markup language (known as "`wiki` markup"), and often edited with the help of a rich-text editor. A `wiki` is run using `wiki` software, otherwise known as a `wiki` engine. There are dozens of different `wiki` engines in use, both stand-alone and part of other software, such as bug tracking systems. The website

`https://en.wikipedia.org/wiki/Wikipedia:How_to_create_a_page`

will tell you how to create a `wiki` page.

### 7.4.9  Wikipedia

As you may know, Wikipedia is a *very large* online encyclopedia—one with nearly 5.5 million entries in English alone.[6] And it is written by the readership. The Wikipedia organization does a fairly careful job of monitoring and vetting the articles, and they are generally of good quality. It is something of an encomium to have a Wikipedia article written about oneself.

You may actually want to consider writing an article for Wikipedia. If, for instance, you work in several complex variables, and you are interested in domains of finite type or boundary orbit accumulation points (both subjects of current intense interest), then you may be disappointed to find that Wikipedia has no article on either of these topics. So you may like to write one. The URL in the preceding paragraph but one will tell you how to do so. The Wikipedia article will *not* identify you as the author, and it *will* allow others to correct and augment your words. It is a collaborative process, and it can be fun.

### 7.4.10  WikiMath

There is also WikiMath, which is a wiki page designed specifically for mathematics and mathematical questions. It has been described as follows:

> This wiki collects tasks and topics from mathematics, including their solutions. This is for everyone who, by himself/herself, feels a need for mathematical help. You can look here for your task. Either you find a solution directly, or you can hope that maybe the next person interested in WikiMath will discuss your project on a new page.

The Wikipedia organization has a number of web pages to orient you towards the project and to get you started as a participant. Among these are

https://en.wikipedia.org/wiki/Wikipedia:Contributing_to_Wikipedia

https://en.wikipedia.org/wiki/Wikipedia:Tutorial

https://en.wikipedia.org/wiki/Wikipedia:Articles_for_Creation

https://en.wikipedia.org/wiki/Wikipedia:Your_first_article

The first of these is a discussion of who should contribute to Wikipedia and how they should do it. What are the rules? And what are the modes of behavior?

The second is a tutorial on how to create an article for Wikipedia.

The third is a discussion of the creative process for Wikipedia.

And the fourth addresses considerations pertaining to your first Wikipedia article.

---

[6]This would be comparable to a 2500-volume hard-copy encyclopedia.

Blogs and chat rooms and `wikis` and related utilities have expanded our ability to communicate with people all over the world. They have augmented the already exploding activity of mathematical collaboration. They are a significant new part of life.

## 7.5  Facebook, Twitter, Instagram, and the Like

### 7.5.1  Social Media

The concept of social media has become quite prominent in the modern world. They foster social relationships, promote events, and advertise ideas; they have played a signficant role in helping people meet mates, and also in helping people to re-gain contact with others whom they haven't seen or communicated with in many years. `Facebook`, `Twitter`, and other utilities play a major role in advertising. Many public figures have social media pages. Newly elected President of the United States Donald Trump frequently uses `Twitter` to promulgate his ideas and opinions.

Social media are computermediated technologies that allow the creating and sharing of information, ideas, career interests, and other forms of expression via virtual communities and networks. The variety of stand-alone and built-in social media services currently available challenges a succinct definition.

However, there are some common features:

1. Social media are interactive web 2.0 internet-based applications.

2. User-generated content, such as text posts or comments, digital photos or videos, and data generated through all online interactions are the lifeblood of social media.

3. Users create service-specific profiles for the website or app that are designed and maintained by the social media organization. These profiles enable users to determine who they are talking to and what they are talking about.

4. Social media facilitate the development of online social networks by connecting a user's profile with those of other individuals and/or groups.

Social media use web-based and mobile technologies on smartphones and tablet computers to create highly interactive platforms through which individuals, communities, and organizations can share, cocreate, discuss, and modify user-generated content or premade content posted online. They introduce substantial and pervasive changes to communication between businesses, organizations, communities, and individuals. Social media change the way businesses, organizations, communities, and individuals interact. Social media change the way individuals and large organizations communicate. These changes are the focus of the emerging field of technoself studies. In America, a survey reported 84% of adolescents have a `Facebook` account. See, `http://time.com/3815154/facebook-teens-pew/` for more details.

## 7.5.2   Social Media and Mathematics

What role can social media play in mathematics? Mathematician Michael Jury was perhaps the first person ever to use social media to prove a theorem. How could he have done this?

`Facebook`, for instance, makes it very easy to communicate with a group of friends—even a very broad group. So Mike used `Facebook` to send a message to a large group of mathematical friends telling of a place where he was stuck in a problem that he was working on. Within the same day he had an answer to his question. He solved the problem and wrote a nice paper.

Certainly it would be possible to send a `Tweet` (a `Twitter` message) announcing that you have proved a nice new theorem. This is not currently the default way to announce a new result. Instead people post their work on `arXiv`, they submit a research announcement to *Research Announcements of the AMS*, or they give a talk at a conference. Or they might send around a mass email. But the world is changing around us and social media may eventually play a more prominent role in mathematicians' lives.

What are some of the things that social media could do for mathematics? Here are some partial answers:

- A social media utility could be used to augment a mathematics class. One could post anecdotes about famous mathematician, bits of mathematical trivia, interesting math facts, challenge problems, and the like. One caution is that it is dangerous territory to make only some of your students your "Friends." Avoid that temptation.

- Social media could be used by mathematics professors to remind students about upcoming tests and other class events (review sessions, films, special presentations, and the like).

- Social media could be used quite effectively to announce and promote upcoming talks, upcoming conferences, new workshops, and other mathematical events.

- One could easily post emendations and errata to lectures on a social media site.

- One could, in principle, use `Facebook` as a tool to promote collaboration. I frankly do not know anyone who does this. Chat rooms would perhaps be more appropriate. Most people that I know collaborate using email—just sending drafts back and forth as attachments.

- One could use social media to announce a current problem in the works and solicit opinions. People could be encouraged to contribute ideas, or even to become more formal collaborators.

- One could start a `Facebook` group to develop ideas about a specific problem.

- `MathType` is a software application created by Design Science that allows the creation of mathematical notation for inclusion in desktop and web applications. One could use `MathType` in your `Facebook` messages.

- One could use `YouTube` to post videos of yourself talking about your latest result. Several journals are now encouraging authors to create videos to go with their published research papers.

- `Mendeley` is a desktop and web program produced by Elsevier for managing and sharing research papers, discovering research data and collaborating online. It combines `Mendeley Desktop`, a `pdf` and reference management application (available for `Windows`, `macOS` and `Linux`) and `Mendeley` for `Android` and `iOS`, with `Mendeley web`, an online social network for researchers.

- `Polymath`, `Stack Exchange`, and `MathOverflow` are social media designed specifically for generating and encouraging large-scale mathematical collaboration. There have already been several significant research problems solved by people working with these utilities.

- One can set up a trackable profile using `Google Scholar` (referenced elsewhere in this book).

- One could use a combination of social media platforms to draw attention to new work by `Tweeting` a link and making a `Facebook` post to your latest paper on `arXiv`.

The website `https://www.researchgate.net/` is a useful tool for scholars. It helps you to keep plugged in to what is going on in your field. It is helpful for disseminating ideas, and also for locating other scholars who share your interests. The website finds papers with similar references and then helps the authors to hook up. The website `www.academia.edu` is similar. `ResearchGate` is a social networking site for scientists and researchers that is used to share papers, ask and answer questions, and find collaborators. According to a study by *Nature* and an article in *Times Higher Education*, it is the largest academic social network in terms of active users, although other services have more registered users and more recent data suggests that almost as many academics have Google Scholar profiles.

People that wish to use the `ResearchGate` site need to have an email address at a recognized institution or to be manually confirmed as a published researcher in order to sign up for an account. Members of the site each have a user profile and can upload research output including papers, data, chapters, negative results, patents, research proposals, methods, presentations, and software source code. Users may also follow the activities of other users and engage in discussions with them. Users are also able to block interactions with other users.

Mathematicians tend to stick to the utilities to which they have become inured. But certainly `ResearchGate` is a tool that we could learn to use and to benefit from.

## 7.5.3   Social Media and Blogs

Social media are often confused with blogs and other Internet utilities.  The distinguishing feature of genuine social media are these:

1. **User Accounts.** Social media require visitors to create unique user accounts in order to interact with other users on each platform. You cannot really share information or interact with others online without doing it through a user account.

2. **Profile Pages.** Social media offer users the ability to curate their own profile page on which to list personal information. This often includes a photo, bio, contact information, website, feed, etc.

3. **Friends, Followers, Groups, Hashtags and so on.** Individuals use their accounts to connect with other users. Here friends are people who know each other, followers do not necessarily know those whom they follow, and groups are associations of people with common interests.  A hashtag is a type of metadata tag used on social network and microblogging services, allowing users to apply dynamic, user-generated tagging that makes it possible for others to easily find messages with a specific theme or content.

4. **Curated Feeds.** When users connect with other users on social media, they're basically saying, "I want to get information from these people." That information is updated for them in real time via their news feed.

   If a site or an app allows you to post absolutely anything, with or without a user account, then it is social. It could be a simple text-based message, a photo upload, a `YouTube` video, a link to an article, or anything else.

5. **User Experience Personalization.**  Social media sites usually give users the flexibility to configure their user settings, customize their profiles to look a specific way, organize their friends or followers, manage the information they see in their news feeds, and even give feedback on what they do or do not want to see.

6. **Notifications.** Any site or app that notifies users about specific information is definitely playing the social media game. Users have control over these notifications and can choose to receive the types of notifications that they want.

7. **Direct Feedback Loop: The Ability to React and Interact in Real Time.** Two of the most common ways we interact on social media are via buttons that represent a "like" or a "dislike" or other reactive feeling plus comment sections where we can share our thoughts.

8. **Review, Rating, or Voting Systems.** Besides liking and commenting, lots of social media sites and apps rely on the collective effort of the community to review, rate, and vote on information that they know about

or have used. `Facebook` allows direct ratings of businesses, etc. through pages. Think of your favorite shopping sites or movie review sites that use this social media feature.

### 7.5.4 Variants of Social Media

There are philosophically related communities on the web that do not exactly fit the moniker "social media." An example is `reddit`. Reddit is a social news aggregation, web content rating, and discussion website. `Reddit`'s registered community members can submit content, such as text posts or direct links. Registered users can then vote submissions up or down to organize the posts and determine their position on the site's pages. The submissions with the most positive votes appear on the front page or the top of a category. Content entries are organized by areas of interest called "subreddits." The sub`reddit` topics include news, science, gaming, movies, music, books, fitness, food, and image-sharing, among many others.

It is possible that `Facebook` and `Twitter` are transient phenomena. But the concept of social media has become quite prominent in modern society. There are currently 1.79 billion `Facebook` users and 313 million `Twitter` users. See the websites `https://www.statista.com/statistics/264810/number-of-monthly-active-facebook-users-worldwide` and `https://www.statista.com/statistics/282087/number-of-monthly-active-twitter-users` for more detail about these numbers. This represents a substantial fraction of the world's population (which is 7.4 billion). While the landscape may change in twenty years, it seems clear that some form of social media will be with us for a good long time.

There are many people who spend several hours everyday recording all the details of their lives on social media—and this is accompanied by copious photographs and other graphics. It is too soon to tell what role social media may play in the mathematical sciences. But much potential exists.

## 7.6 Print-on-Demand Books

### 7.6.1 Instantly Produced Books

In the old days—fifty years ago, let us say—the production of a book was very formulaic. First you wrote the book by hand on paper. Then you had it typed up. You submitted a hard copy to the publisher. Once the book was reviewed and accepted by the publisher, then it was typeset in cold type. Then proof sheets were printed. The final-form proof sheets were photographically "shot" and then printed on copper plates with a special acid-resistant ink. At the next step, photo engravers used sulphuric acid to etch the copper plates which were then used as the printing plates on a high-speed printing press. And a print run had to be at least 1000 units in order to be cost effective. Printing just a few copies of a book was virtually infeasible. No more.

Nowadays print-on-demand is both feasible and cost effective. Because of electronic media, there is no longer a notion of a book "going out of print." The electronic data for a book—stored on a hard drive or a flash drive or a tape—is always there. And backed up on remote devices as well. There are numerous companies—including Book1One, Xulon Publishers, and altagraphics—which can produce print-on-demand books for you.

## 7.6.2  The Espresso Book Machine

The Espresso Book Machine (EBM) is a print-on-demand machine that prints, collates, covers, and binds a single book in a few minutes. The EBM is small enough to fit in a retail bookstore or small library room, and as such it is targeted at retail and library markets. The EBM can potentially allow readers to obtain any book title, even books that are out of print. The machine takes a `*.pdf` file for input and prints, binds, and trims the reader's selection as a paperback book.

Jason Epstein gave a series of lectures in 1999 about his experiences in publishing. Epstein mentioned in his talks that a future was possible in which customers would be able to print an out-of-stock title on the spot, if a book-printing machine could be made that would fit in a store. He founded 3BillionBooks with Michael Smolens, a Long Island entrepreneur from Russia, and Thor Sigvaldason, a consultant at Price Waterhouse Coopers. At the time, Jeff Marsh, a St. Louis engineer and inventor, had already constructed a prototype book printer that could both photocopy and bind. Marsh was working on this project for Harvey Ross, who held a patent for such a machine. Peter Zelchenko, a Chicago-based technologist and a partner of Ross in a related patent effort, worked with Marsh to prove the concept and also helped bring Marsh and other players together with several venture interests.

Ultimately Epstein, together with Dane Neller, former President and CEO of Dean and Deluca, licensed Marsh's invention and founded On Demand Books. The first Espresso Book Machine was installed and demonstrated on June 21, 2007, at the New York Public Library's Science, Industry, and Business Library. For a month, the public was allowed to test the machine by printing free copies of public domain titles provided by the Open Content Alliance, a non-profit organization with a database of over 200,000 titles.

The direct-to-consumer model supported by the Espresso Book Machine eliminates the need for shipping, warehousing, returns, and pulping of unsold books; it allows simultaneous global availability of millions of new and backlist titles.

Unfortunately the Espresso Book Machine costs about $150,000. Or you can lease it for $5,000 per month. So this is out of reach for most people. But a number of bookstores and libraries have one, and let their customers use it for a modest fee. It is possible, at least in principle, to produce a hard copy of a `Google` digitized book (which is, by definition, Open Access) for about $8.

Now it is conceivable to have your working seminar at University X put together a book gathering together the thoughts that you have been developing for the past few years and have several copies printed up for use by the group (and for the students as well).

Remember that, as soon as you write something, then it is automatically copyrighted to you. So you need not worry about protecting your book once it is printed.

### 7.6.3 CreateSpace

Today `Amazon` is the world's largest book seller—and seller of everything else as well. `Amazon` has revolutionized the book business in many ways. `CreateSpace` is an artifact of `Amazon` that allows you to create your own electronic book to be posted on `Amazon`. And `Amazon` can produce the book in hard copy as well.

It has been six years since Amazon acquired `CreateSpace`, an on-demand publishing platform, and almost four years since they announced the free online setup for self-publishing. While four years seems like a long time in our fast-paced world, self-publishing still has not reached the mass audience. Even the biggest social media gurus still take the traditional route, only choosing to self-publish when they've been rejected by mainstream publishing houses.

### 7.6.4 Print-on-Demand Publishing

The truth is that print-on-demand publishing is the fastest, most profitable, and easiest way to get your written thoughts out there. Today, self-published books are even distributed to traditional outlets like Barnes & Noble and academic libraries.

Of course self-publishing means you do not get the marketing resources that come with a traditional publishing deal, but in our world of social media, that can be easily fixed. So if self publishing is so easy, why do we not see more authors using it? Most people are simply not aware of the low barrier to entry. This could be the wave of the future.

# Chapter 8

# Closing Thoughts

*Of all those arts in which the wise excel,*
*Nature's chief masterpiece is writing well.*

John Sheffield, Duke of Buckingham and Normanby
Essay on Poetry [1682]

*Great prizes are reader interest and understanding; all else is secondary. Graceful prose, imagery, wit, even orthography and grammar are only means to more important ends. This observation makes writing and reading more of a colloquy and less a lonely or isolating business.*

from the dust jacket of *Mathematical Writing* [KnLR]

*England has forty-two religions and two sauces.*

Voltaire

*A writer needs three things, experience, observation, and imagination, any two of which, at times any one of which, can supply the lack of the others.*

William Faulkner

*Isaac Newton invented his theory of gravity when he was 21. I'm 32, and I just found out that Garfield and Heathcliff are two different cats.*

Anon.

*Anything that helps communication is good. Anything that hurts is bad.*

Paul Halmos

## 8.1 Why Is Writing Important?

### 8.1.1 The Case for Good Writing

The case for writing, indeed for good writing, has been made throughout this book. Writing is our tool for communicating our ideas, and for leaving a legacy for future generations. One of the marvels of genuinely outstanding writing is its longevity. In many ways, the writings of Herodotus, of Descartes, of Plato, or of Faulkner are as vibrant and important today as when they were first penned.

Writing at the very highest level is often painstaking and tedious. A good author can spend an entire day agonizing over a word, a comma, or a phrase. He/she will revise the work mercilessly. For the working mathematician, I am not recommending this sort of writing. Do it if you like; but this level of precision and artistry is not what our profession either demands or needs. In fact the sort of clear, cogent, precise writing that I am promoting here requires little more effort than lousy writing requires. Like the ability to scuba dive, the ability to write well is in truth a matter of becoming conversant with the basic principles and then practicing. Once you become comfortable with the process, then writing becomes less of a chore and more of a pleasurable pastime. It allows you to view your written work as an accomplishment to be proud of, rather than another agony that you have slogged your way through.

## 8.1.2   Do We Know How to Write?

We all grow up speaking English (or some other native language). After a while, we convince ourselves that we are able to express our thoughts verbally—regardless of our technical facility with grammar, usage, and syntax. As we grow older, a corollary of such reasoning is that we all think that we know how to write. A result of this process is that it is more difficult to teach people to write (and, in turn, for them to learn to write) than it is to teach people calculus. When a student has his/her calculus work marked incorrect, then he/she is inclined to say "Apparently I don't know how to do this kind of problem. I'd better get some help." But when a student has his/her writing marked up and criticized, then he/she is liable to go to the instructor and say "Well, just what is it that you want?"

Learning to write well is a yoga; it is a manner of being trained in self-criticism and self-instruction. Fields Medalist Enrico Bombieri has observed to me that his artistic activities, particularly his painting, have helped him to see things more clearly, and in greater detail. Just so, learning to write well will sharpen your thoughts, develop your skills at ratiocination, and help you to communicate more effectively.

Developing an ability to write cogently and incisively will give you an appreciation of the writing, and the thinking, of others. And you will learn from their writing—both what to do and what not to do. It will add a new dimension to your life. I hope that it is a happy one.

# Bibliography

[Ad1]   *The Adams Cover Letter Almanac and Disc*, Adams Media Corpo-
        ration, Holbrook, Massachusetts, 1996.

[Ad2]   *The Adams Jobs Almanac*, Adams Media Corporation, Holbrook,
        Massachusetts, 1996.

[Ad3]   *The Adams Resumés Almanac and Disc*, Adams Media Corporation,
        Holbrook, Massachusetts, 1996.

[Ati]   M. Atiyah, et al., Responses to "Theoretical Mathematics: Toward a
        Cultural Synthesis of Mathematics and Theoretical Physics", by A.
        Jaffee and F. Quinn, *Bulletin of the AMS* 30(1994), 178–207.

[BM]    G. Birkhoff and S. MacLane, *A Survey of Modern Algebra*, MacMil-
        lan, New York, 1941.

[Bae]   J. Baez, Math Blogs, *Notices of the AMS* 57(2010), 333.

[Bal]   P. Ball, Leading mathematician launches arXiv "overlay" journal,
        Nature vol. 526, October 1, 2015.

[Blo]   S. Bloch, Review of *Étale Cohomology* by J. S. Milne, *Bulletin of the
        AMS*, new series, 4(1981), 235-239.

[Caj]   F. Cajori, *A History of Mathematical Notations*, The Open Court
        Publishing Company, Chicago, IL, 1928–29.

[Chi]   *A Manual of Style*, 14th Edition, (Chicago: The University of Chicago
        Press, 1993).

[Dub]   E. Dubinsky, A. Schoenfeld, J. Kaput, eds. and T. Dick (managing
        ed.), *Research in Collegiate Mathematics Education, I*, The AMS in
        cooperation with the MAA, Providence, Rhode Island, 1994.

[DS]    N. Dunford and J. Schwartz, *Linear Operators*, Wiley Interscience,
        New York, 1958-1971, 1988.

[Dup]   L. Dupré, *Bugs in Writing*, Addison-Wesley, Reading, Massachusetts,
        1995.

[Dys]     F. Dyson, Missed Opportunities, *Bulletin of the AMS* 78(1972), 635-652.

[Ewi]     J. Ewing, Where are journals headed? Why we should worry about author-pay, *Notices of the AMS* 55(2008), 381–382.

[Fow]     H. W. Fowler, *Modern English Usage*, Oxford University Press, Oxford, 1965.

[Fra]     M. Frank, *Modern English: A Practical Reference Guide*, 2nd edition, Regents/Prentice-Hall, Englewood Cliffs, New Jersey, 1993.

[Gil]     L. Gillman, *Writing Mathematics Well: A Manual for Authors*, The Mathematical Association of America, Washington, D.C., 1987.

[Gle]     J. Gleick, *Chaos: Making a New Science*, Viking, New York, 1987.

[Gos]     M. Goossens, F. Mittelbach, and Alexander Samarin, *The LaTeX Companion*, Addison-Wesley, Reading, MA, 1994.

[Gra1]    G. Grätzer, *First Steps in LaTeX*, Birkhäuser, Boston, 1999.

[Gra2]    G. Grätzer, *Practical LaTeX*, Springer, New York, 2014.

[Gra3]    G. Grätzer, *More Math into LaTeX*, 5th ed., Springer, New York, 2016.

[GH]      P. Griffiths and J. Harris, *Principles of Algebraic Geometry*, John Wiley and Sons, New York, 1978.

[Hig]     N. J. Higham, *Handbook of Writing for the Mathematical Sciences*, SIAM, Philadelphia, Pennsylvania, 1993.

[Hor]     J. Horgan, The Death of Proof, *Scientific American*, October, 1993.

[JQ]      A. Jaffe and F. Quinn, "Theoretical Mathematics": Toward a cultural synthesis of mathematics and theoretical physics, *Bulletin of the AMS* 29(1993), 1-13.

[Ker]     N. Kerzman, Hölder and $L^p$ estimates for solutions of $\overline{\partial}u = f$ in strongly pseudoconvex domains, *Comm. Pure Appl. Math.* 24(1971), 301–379.

[Kli]     W. Klingenberg, Review of *Affine Differential Geometry* by K. Nomizu and T. Sasaki, *Bulletin of the AMS*, new series, 33(1996), 75-76.

[Kn]      D. E. Knuth, Mathematical typography, *Bulletin of the AMS* (new series) 1(1979), 337-372.

[KnT]     D. E. Knuth, *The TeXBook*, Addison-Wesley, Reading, MA, 1986.

[KnLR]  D. E. Knuth, T. Larrabee, and P. M. Roberts, *Mathematical Writing*, Mathematical Notes No. 14, The Mathematical Association of America, Washington, D.C., 1989.

[Kr1]  S. G. Krantz, *Real Analysis and Foundations*, CRC Press, Boca Raton, Florida, 1991.

[Kr2]  S. G. Krantz, *How to Teach Mathematics*, American Mathematical Society, Providence, Rhode Island, 1992.

[Kur]  K. Kuratowski, *Introduction to Set Theory and Topology*, Addison-Wesley, Reading, Massachusetts, 1961.

[Lam]  L. Lamport, *LaTeX: A Document Preparation System. User's Guide and Manual.*, $2^{nd}$ Ed., Addison-Wesley, Reading, Massachusetts, 1994.

[Lan]  S. Lang, Mordell's review, Siegel's letter to Mordell, Diophantine geometry, and $20^{th}$ century mathematics, *Notices of the AMS* 42(1995), 339-350.

[Lip]  J. Lipman, Review of *Principles of Algebraic Geometry* by P. Griffiths and J. Harris, *Bulletin of the AMS*, new series, 2(1980), 197-200.

[McL1]  M. McLuhan and Q. Fiore, *The Medium is the Massage*, Random House, New York, 1967.

[McL2]  M. McLuhan and B. Powers, *The Global Village*, Oxford University Press, New York, 1989.

[MW]  *Merriam-Webster's Dictionary of English Usage*, Merriam-Webster, Inc., Springfield, Massachusetts, 1994.

[Mit]  M. Mitchell, *Gone with the Wind*, Macmillan, New York, 1936.

[MGBCR]  F. Mittelbach, M. Goossens, J. Braams, D. Carlisle, C. Rowley, *The LaTeX Companion*, 2nd ed., Addison-Wesley, Reading, MA, 2004.

[Moo]  G. E. Moore, *Ethics*, Holt and Company, New York, 1912.

[Mor]  L. Mordell, Review of *Diophantine Geometry* by S. Lang, *Bulletin of the AMS* 70(1962), 491-498.

[New]  E. Newman, *Strictly Speaking*, Warner Books, New York, 1974.

[NoS]  K. Nomizu and T. Sasaki, *Book Review by Klingenberg*, Letter to the Editor, *Notices of the AMS* 43(1996), 655-656.

[Oke]  A. Okerson, Whose article is it anyway?, *Notices of the AMS* 43(1996), 8-12.

[PS]     H. O. Peitgen and D. Saupe, *The Science of Fractal Images*, Springer-Verlag, Berlin, 1989.

[Pub]    Public access to NIH research made law, `www.sciencecodex.com`.

[Saf]    W. Safire, *New York Times*, January 8, 1989, Section 6, p. 12.

[SaK]    S. Sawyer and S. G. Krantz, *A TEX Primer for Scientists*, CRC Press, Boca Raton, Florida, 1995.

[SG]     M. E. Skillin, R. M. Gay, et al., *Words into Type*, Prentice-Hall, Englewood Cliffs, New Jersey, 1994.

[SPI]    M. Spivak, *The Joy of TEX*, American Mathematical Society, Providence, RI, 1990.

[Ste]    N. Steenrod, et al., *How to Write Mathematics*, American Mathematical Society, Providence, Rhode Island, 1973.

[SW]     W. Strunk and W. White, *The Elements of Style*, 3$^{\text{rd}}$ Edition, Macmillan, New York, 1979.

[Sub]    P. Suber, *Open Access*, MIT Press, Cambridge, MA, 2012.

[Swa]    E. Swanson, *Mathematics into Type*, revised edition, the AMS, Providence, Rhode Island, 1979.

[Tuf1]   E. Tufte, *The Visual Display of Quantitative Information*, Graphics Press, Cheshire, Connecticut, 1983.

[Tuf2]   E. Tufte, *Envisioning Information*, Graphics Press, Cheshire, Connecticut, 1990.

[VanL]   M. C. van Leunen, *Handbook for Scholars*, revised ed., Oxford University Press, New York, 1992.

[Vas]    B. Vastag, NIH Launches PubMed Central, *Journal of the National Cancer Institute* 92(2000), 374–374.

[Wei]    A. Weil, *Basic Number Theory*, Die Grundlehren der mathematischen Wissenschaften, Band 144, Springer-Verlag New York, Inc., New York, 1967.

[Wer]    J. Wermer, Letter to the Editor of the *Notices of the AMS* 42(1995), 5.

[WR]     A. N. Whitehead and B. Russell, *Principia Mathematica*, Second Edition, Cambridge Univ. Press, Cambridge, 1950.

# Index